61th International Mathematical Olympiad
IMO 2020 | Saint Petersburg – Russia

Educational Collection *Magna-Scientia*

61th International Mathematical Olympiad

IMO 2020 | Saint Petersburg – Russia

Michael Angel C. G., Editor

Preface

The International Mathematical Olympiad (IMO) is the World Math Competition for high school students and is held annually in a different country, establishing itself as the most prestigious Math competition that a high school student can aspire to take part. The first IMO was held in 1959 in Romania, with 7 participating countries. Since then, it has gradually expanded to more than 100 countries on 5 continents.

Likewise, the IMO is a great opportunity for students to face original, challenging and interesting math problems; which can be used to measure their level of knowledge to other students around the world. Among the topics covered by the problems we have: Algebra, Combinatorics, Geometry and Number Theory.

On this occasion we make available to the student, a bilingual edition (English-Spanish) of the exam with detailed solutions of the 61th International Mathematical Olympiad held virtually from Saint Petersburg – Russia in September 2020. In addition, an appendix with problem statements from IMO exams between 2010 and 2019 are included.

Finally, I hope with optimism that this book will contribute to the preparation of the high school students in these types of competitions, inviting them to face each problem as a personal challenge before reviewing the respective solution.

Sincerely,

The editor

Contents

In English

Problems

IMO 2020

IMO 2020

61ᵗʰ International Mathematical Olympiad

Saint Petersburg, Russia | September 19 – 28, 2020.

Day 1 (Sept. 21, 2020)

Problem 1 (By Dominik Burek, Poland)
In the interior of a convex quadrilateral $ABCD$, there is a point P such that the following equalities hold:

$$\angle PAD : \angle PBA : \angle DPA = 1 : 2 : 3 = \angle CBP : \angle BAP : \angle BPC.$$

Show that the following three lines intersect at one point: the internal bisectors of the angles ADP and PCB and the perpendicular bisector of segment AB.

Problem 2 (By Stijn Cambie, Belgium)
Given the real numbers a, b, c, d such that $a \geq b \geq c \geq d > 0$ and $a + b + c + d = 1$. Show that,

$$(a + 2b + 3c + 4d)\, a^a \cdot b^b \cdot c^c \cdot d^d < 1.$$

Problem 3 (By Carl Schildkraut and Milan Haiman, Hungary)
There are $4n$ pebbles with masses $1, 2, 3, \ldots, 4n$. Each of the pebbles is painted in one of n colors, and there are 4 pebbles of each color. Show that the pebbles can be divided into two piles of equal total weight so that each pile contains two pebbles of each color.

Problem 4 (By Tejaswi Navrinarekallu, India)

Given an integer $n > 1$. There are n^2 cable car stations on the slope of a mountain at different altitudes. Each of the two cable car companies A and B owns k cable cars. Each cable car makes a regular non-stop transfer from one station to a higher one. It is known that k transfers from company A begin at k different stations and end at k different stations; in addition, transfers that start at the top also end at the top. The same conditions are met for company B. We will say that two stations are *connected* by a cable car company if it is possible to get from the lower station to the higher one using one or more transfers from this company (other transfers between stations are prohibited). Find the smallest integer k so that there are two stations connected by both companies.

Problem 5 (By Oleg Košik, Estonia)

There are $n > 1$ cards in a deck, each of which contains a positive integer. It turned out that, for any two cards, the arithmetic mean of the numbers written on them is equal to the geometric mean of the numbers written on the cards of a given set consisting of one or more cards. For what n are all the numbers written on the cards equal?

Problem 6 (By Ting-Feng Lin and Hung-Hsun Hans Yu, Taiwan)

Show that there is a positive constant c for which the following statement holds:

Let S be a set of $n > 1$ points in the plane, in which the distance between any two points is not less than 1. Then there is a line ℓ that separates the set S such that the distance from any point of S to ℓ is not less than $cn^{-1/3}$.

(A line ℓ *separates* the set of points S if it cuts some segment whose ends belong to S.)

Note. Weaker results obtained by replacing $cn^{-1/3}$ with $cn^{-1/\alpha}$ may be awarded points depending on the value of the constant $\alpha > 1/3$.

Solution
IMO 2020

Problem 1 (By Dominik Burek, Poland)

In the interior of a convex quadrilateral ABCD, there is a point P such that the following equalities hold:

$$\angle PAD : \angle PBA : \angle DPA = 1 : 2 : 3 = \angle CBP : \angle BAP : \angle BPC.$$

Show that the following three lines intersect at one point: the internal bisectors of the angles ADP and PCB and the perpendicular bisector of segment AB.

Solution

Put $\varphi = \angle PAD$ and $\psi = \angle CBP$; then $\angle PBA = 2\varphi$, $\angle DPA = 3\varphi$, $BAP = 2\psi$ and $\angle BPC = 3\psi$ (see Fig. 1). Let X be a point on segment AD such that $\angle XPA = \varphi$. We have

$$\angle PXD = \angle PAX + \angle XPA = 2\varphi = \angle DPA - \angle XPA = \angle DPX.$$

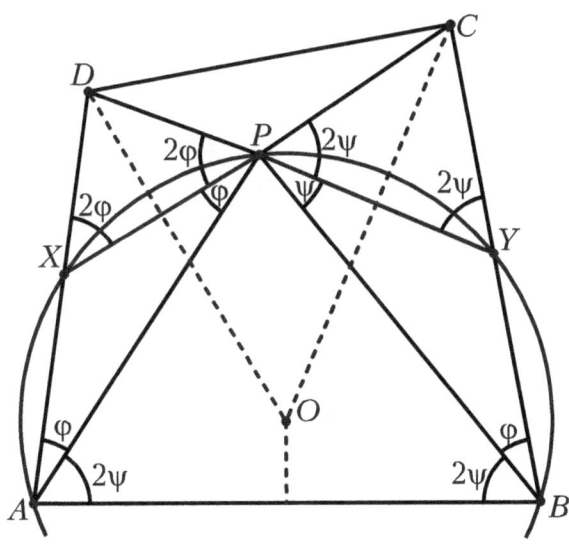

Fig. 1

Then the triangle DPX is isosceles with lateral sides $DX = DP$, therefore the bisector of the angle ADP is the perpendicular bisector of XP. Similarly, if Y is a point on BC such that $\angle BPY = \psi$, then the bisector of angle PCB is the perpendicular bisector of PY. So, we need to show that the perpendicular bisectors of the segments XP, PY and AB meet at a point.

It is observed that $\angle AXP = 180° - \angle PXD = 180° - 2\varphi = 180° - \angle PBA$. Therefore, the quadrilateral $AXPB$ is inscribed, and X is on the circumscribed circle of triangle APB. Similarly, Y lies on the circumscribed circle of triangle APB. Therefore, the perpendicular bisectors of the segments XP, PY, and AB pass through the center of the circle $(ABYPX)$. This completes the proof.

Problem 2 (By Stijn Cambie, Belgium)

Given the real numbers a, b, c, d such that $a \geq b \geq c \geq d > 0$ and $a + b + c + d = 1$. Show that,

$$(a + 2b + 3c + 4d)\, a^a \cdot b^b \cdot c^c \cdot d^d < 1.$$

Solution

We apply the generalized mean (weighted) inequality to the positive reals $x_1, \ldots, x_n$ and $\lambda_1, \ldots, \lambda_n$ such that $\lambda_1 + \cdots + \lambda_n = 1$,

$$x_1^{\lambda_1} \cdot x_2^{\lambda_2} \cdot \ldots \cdot x_n^{\lambda_n} \leq \lambda_1 x_1 + \lambda_2 x_2 + \cdots + \lambda_n x_n.$$

We have

$$a^a \cdot b^b \cdot c^c \cdot d^d \leq a \cdot a + b \cdot b + c \cdot c + d \cdot d = a^2 + b^2 + c^2 + d^2,$$

therefore, it suffices to prove that

$$(a + 2b + 3c + 4d)\,(a^2 + b^2 + c^2 + d^2) < 1 = (a + b + c + d)^3.$$

The next inequality

$$(a + b + c + d)^3 \geq (a + 2b + 3c + 4d)\,(a^2 + b^2 + c^2 + d^2)$$

can be proved, for example, as follows. We have

$$(a + b + c + d)^3 > a^2(a + 3b + 3c + 3d) + b^2(3a + b + 3c + 3d) +$$
$$+ c^2(3a + 3b + c + 3d) + d^2(3a + 3b + 3c + d), (*)$$

since all the terms on the right side are explicitly present after expanding the brackets on the left side. Since each of the expressions $a + 3b + 3c + 3d$, $3a + b + 3c + 3d$, $3a + 3b + c + 3d$, $3a + 3b + 3c + d$ is not less than $a + 2b + 3c + 4d$, the right side in $(*)$ is not less than $(a + 2b + 3c + 4d)\,(a^2 + b^2 + c^2 + d^2)$.

19

Problem 3 (By Carl Schildkraut and Milan Haiman, Hungary)
There are 4n pebbles with masses $1, 2, 3, \ldots, 4n$. Each of the pebbles is painted in one of n colors, and there are 4 pebbles of each color. Show that the pebbles can be divided into two piles of equal total weight so that each pile contains two pebbles of each color.

Solution
We combine pebbles in pairs with sum of masses $4n + 1$, so that we obtain a set S that consists of $2n$ pairs: $(1, 4n)$, $(2, 4n - 1)$, $(3, 4n - 2)$, ..., $(2n, \; 2n + 1)$. It is sufficient to divide S into two subsets of n pairs so that each subset contains two pebbles of each color. Let us introduce a multigraph G (that is, a graph in which loops and multiple edges are allowed) at n vertices such that each vertex corresponds to a certain color. For each pair of pebbles in set S, draw an edge between the vertices corresponding to the colors of these pebbles. In this case, the degree of each vertex will be 4. Also note that the required division of the pebbles corresponds to coloring the edges of the graph in two colors, *e.g.* blue and red, so that exactly 2 red and 2 blue edges emerge of each vertex.

To complete the solution, it is enough to match a color to each connected component G^c of graph G. Since the degrees of all vertices are even, there is a Euler path P in G^c (that is, a cycle that passes through of each edge of G^c exactly once). Note that the number of edges in P is even (it is equal to twice the number of vertices in G^c). Therefore, all edges can be colored red and blue so that two adjacent edges of P have different colors (it can be traversed along P and color the edges, alternating red and blue). Therefore, the same number of red and blue edges emerge from each vertex in G^c, as needed.

Problem 4 (By Tejaswi Navrinarekallu, India)

Given an integer $n > 1$. There are n^2 cable car stations on the slope of a mountain at different altitudes. Each of the two cable car companies A and B owns k cable cars. Each cable car makes a regular non-stop transfer from one station to a higher one. It is known that k transfers from company A begin at k different stations and end at k different stations; in addition, transfers that start at the top also end at the top. The same conditions are met for company B. We will say that two stations are "connected" by a cable car company if it is possible to get from the lower station to the higher one using one or more transfers from this company (other transfers between stations are prohibited). Find the smallest integer k so that there are two stations connected by both companies.

Solution

Answer. $k = n^2 - n - 1$.

First, we show that for any $k \leq n^2 - n$, there may not be a pair of stations connected by each of the two companies. Suffice it to give an example for $k = n^2 - n$.

Suppose that company A has transfers $(i, i + 1)$, where i is not divisible by n, and that company B has transfers of the form $(i, i + n)$, where $1 \leq i \leq n^2 - n$. Then, the numbers i, j of the stations connected by company A are such that $|i - j| \leq n - 1$. It is clear that these stations are not connected by company B.

Note that all final stations of $A -$ transfers are different, therefore there are $n^2 - k = n - 1$ stations $a_1, \ldots, a_{n-1}$, which are not final for $A -$ transfers. From station a_1 we will go up for $A -$ transfers as high as possible. This route gives the maximum $A -$ chain of stations. Similarly, each of the stations $a_2, \ldots, a_{n-1}$ determines the maximum $A -$ chain (possibly consisting of a single station). It is clear that each of the n^2 stations are included in one of the maximum $A -$ chains (since for each station the path down by $A -$ transfers to one of the

21

stations $a_1, \ldots, a_{n-1}$ is uniquely determined). Similarly, we can show that each station belongs to one of the $n - 1$ maximum $B -$ chains. Note that one of the maximum $A -$ chains contain at least $n^2 / (n - 1) > n$ stations, and at least two of these stations belong to the same maximum $B -$ chain. Then, the two stations found are the desired ones.

Problem 5 (By Oleg Košik, Estonia)

There are $n > 1$ cards in a deck, each of which contains a positive integer. It turned out that, for any two cards, the arithmetic mean of the numbers written on them is equal to the geometric mean of the numbers written on the cards of a given set consisting of one or more cards. For what n are all the numbers written on the cards equal?

Solution

Answer. For all n.

Suppose that not all numbers are equal, and $b_1 > b_2 > \ldots > b_k$ are all different numbers among the numbers on the cards, $k > 2$ (so that the set of numbers on the cards consists of several copies of the b_1 numbers, several copies of b_2 numbers, etc.). If $d = \gcd(b_1, \ldots, b_k) > 1$, then dividing all the numbers on the cards by d, we arrive at an example that satisfies the condition and in which the gcd of all the numbers on the cards is 1. Then, in what follows we will assume that $\gcd(b_1, \ldots, b_k) = 1$. Let p be a prime divisor of b_1 (it exists, since $b_1 > 1$). Since $\gcd(b_1, \ldots, b_k) = 1$, there is a minimum $m \geq 2$ such that b_m is not a multiple of p.

The number $q = (b_{m-1} + b_m)/2$ has to be the geometric mean of several b_i, that is, $q^t = c_1 c_2 \ldots c_t$ for some $t \in \mathbb{N}$ and $c_i \in \{b_1, \ldots, b_k\}$. Note that some c_i match one of the numbers $b_1, \ldots, b_{m-1}$, because otherwise $c_1 c_2 \ldots c_t \leq b_m^t < q^t$. Therefore, p is a divisor of some c_i and $c_1 c_2 \ldots c_t$ is divisible by p. Likewise, the expression $q^t = c_1 c_2 \ldots c_t$ reduces to the following equality in integers:

$$(b_{m-1} + b_m)^t = 2^t \cdot c_1 c_2 \ldots c_t.$$

Since the right side is divisible by p, then $b_{m-1} + b_m$ is divisible by p. But on the other hand, by the definition of m, b_{m-1} is divisible by p, but b_m is not. We get a contradiction; then it is impossible that for any $n > 1$ there are different numbers on the cards that satisfy the condition. Finally, for all n the numbers on the cards should be equal.

Problem 6 (By Ting-Feng Lin and Hung-Hsun Hans Yu, Taiwan)

Show that there is a positive constant c for which the following statement holds:

Let S be a set of n > 1 points in the plane, in which the distance between any two points is not less than 1. Then there is a line ℓ that separates the set S such that the distance from any point of S to ℓ is not less than $cn^{-1/3}$.

(A line ℓ "separates" the set of points S if it cuts some segment whose ends belong to S.)

Note. Weaker results obtained by replacing $cn^{-1/3}$ with $cn^{-1/\alpha}$ may be awarded points depending on the value of the constant $\alpha > 1/3$.

Solution

Let us prove the statement required for $c = 1/8$. We put $\delta = 1/8 \cdot n^{-1/3}$. For any line ℓ and any point X, we denote by X_ℓ the projection of X on ℓ; we use a similar notation for sets of points.

Suppose that for some line ℓ the set S_ℓ contains two neighboring points X and Y with a distance $XY = 2d$. Then, the perpendicular bisector of segment XY separates S, and all the points of the set S will be at a distance of at least d from the straight line ℓ. Therefore, if $d \geq \delta$, then the required line has been found. Suppose, on the contrary, that for any line ℓ in the set S_ℓ, the distance between adjacent points is less than 2δ.

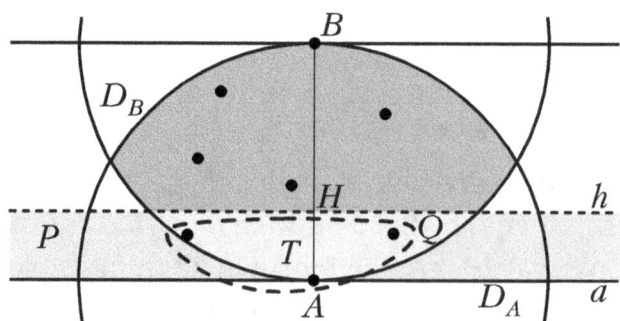

Fig. 2

24

We choose the points $A, B \in S$ with a maximum distance $M = AB$ from each other, so that AB is the diameter of the set S (see Fig. 2); obviously, $M \geq 1$. Let ℓ be the line AB. The set S is at the intersection of two circles D_A and D_B of radius M centered at points A and B. Therefore, projection S_ℓ lies on segment AB. In addition, the points of the set S_ℓ divide the segment AB into at most $n - 1$ parts, the length of each of which is less than 2δ. Thus

$$M < n \cdot 2\delta. \tag{1}$$

On segment AB, we take a point H such that $AH = 1/2$. Let P be a strip between lines a and h that pass through A and H perpendicular to AB (we assume that the limits of the strip P belong to it). We put $T = P \cap S$ and let $t = |T|$. According to our assumption, segment AH does not contain less than $\lceil 1/2 : (2\delta) \rceil$ points of the set S_ℓ, whence

$$t \geq \frac{1}{4\delta}. \tag{2}$$

Note that T is contained in $Q = P \cap D_B$. The set Q is a segment and its projection Q_a is a segment of length

$$2\sqrt{M^2 - \left(M - \frac{1}{2}\right)^2} < 2\sqrt{M}.$$

On the other hand, for any two points $X, Y \in T$, we have $XY \geq 1$ and $X_\ell Y_\ell \leq 1/2$, which means that $X_a Y_a = \sqrt{XY^2 - X_\ell Y_\ell^2} \geq \sqrt{3}/2$. Therefore, the t points that form the set T_a are in an interval of length less than $2\sqrt{M}$, and are separated by at least $\sqrt{3}/2$ from each other. Therefore, $2\sqrt{M} > (t - 1)\sqrt{3}/2$, or

$$t < 1 + \frac{4\sqrt{M}}{\sqrt{3}} < 4\sqrt{M}, \tag{3}$$

since $M \geq 1$.
From the expressions (1), (2) and (3), we obtain

$$\frac{1}{4\delta} \le t < 4\sqrt{M} < 4\sqrt{2n\delta},$$

hence $512n\delta^3 > 1$, which is not true for δ.

Appendix:

Statements
IMO 2010 – 2019

IMO 2019

60th International Mathematical Olympiad

Bath, United Kingdom | July 11 – 22, 2019.

Day 1 (July 16, 2019)

Problem 1 (By Liam Baker, South Africa)

Let $\mathbb{Z}$ be the set of integers. Find all functions $f : \mathbb{Z} \longrightarrow \mathbb{Z}$ such that, for all integers x and y,

$$f(2x) + 2f(y) = f(f(x + y)).$$

Problem 2 (By Anton Trygub, Ukraine)

In triangle ABC, point A_1 is on side BC and point B_1 is on side AC. Let P and Q be points on the segments AA_1 and BB_1, respectively, such that PQ is parallel to AB. And let P_1 be a point on the line PB_1 different from B_1, with B_1 between P and P_1, and $\angle PP_1C = \angle BAC$. Similarly, let Q_1 be a point on the segment QA_1 different from A_1, with A_1 between Q and Q_1, and $\angle CQ_1Q = \angle CBA$. Show that the points P, Q, P_1 and Q_1 are concyclic.

Problem 3 (By Adrian Beker, Croatia)

A social network has 2019 users, where some of them are friends. Whenever user A is a friend of user B, user B is also a friend of user A. Likewise, events of the following type can occur repeatedly, but one at a time:

Three users A, B and C so that A is friends with B and C, but B and C are not friends, they change their friendship relationships so that B and C are now friends, but A is no longer friends with B and C. Other friendship relationships do not change.

Initially there are 1010 users who have 1009 friends each, and there are 1009 users who have 1010 friends each. Show that there is a sequence of this type of events after which each user is a friend of at most one of the other users.

Day 2 (July 17, 2019)

Problem 4 (By Gabriel Chicas Reyes, El Salvador)
Find all pairs of positive integers (k, n) that satisfy the equality

$$k! = (2^n - 1)(2^n - 2)(2^n - 4)\ldots(2^n - 2^{n-1}). \qquad (*)$$

Problem 5 (By David Altizio, United States)
The Bank of Bath issues coins with an H on one side and a T on the other. Harry has n such coins, arranged in a row from left to right. Likewise, he repeatedly carries out the following operation: if there are exactly k $(k > 0)$ coins showing an H, Harry flips the k-th coin counting from the left; otherwise, all coins show a T and the process stops. For example, if $n = 3$ and the initial setting is THT, the process would be $THT \rightarrow HHT \rightarrow HTT \rightarrow TTT$, stopping after three operations.
(a) Show that for whatever initial configuration Harry has, the process stops after a finite number of operations.
(b) For each initial configuration K, let $L(K)$ be the number of operations that are performed until the process stops. For example, $L(THT) = 3$ and $L(TTT) = 0$. Find the average value of $L(K)$ of all 2^n possible initial configurations of K.

Problem 6 (By Anant Mudgal, India)
Let I be the incenter of the acute triangle ABC such that $AB \neq AC$. And let ω be the inscribed circle of triangle ABC, which is tangent to sides BC, CA, and AB at points D, E, and F, respectively. Likewise, the line that passes through D and perpendicular to EF, intersects ω

30

again at R. Furthermore, the segment AR intersects ω again at P. And the circumscribed circles of triangles PCE and PBF intersect again at Q. Show that lines DI and PQ intersect on the line through A and perpendicular to AI.

IMO 2018

59th International Mathematical Olympiad

Cluj-Napoca, Romania | July 03 – 14, 2018.

Day 1 (July 09, 2018)

Problem 1 (By S. Brazitikos, E. Psychas and M. Sarantis, Greece)
Let Γ be the circumscribed circle of the acute triangle ABC. Points D and E are on segments AB and AC, respectively, and are such that $AD = AE$. The perpendicular bisectors of BD and CE intersect the minor arcs AB and AC of Γ at points F and G, respectively. Show that the lines DE and FG are parallel (or are the same line).

Problem 2 (By Patrik Bak, Slovakia)
Find all integers $n \geq 3$ for which there are real numbers $a_1, a_2, \ldots, a_{n+2}$, such that $a_{n+1} = a_1$ and $a_{n+2} = a_2$, if $a_i \cdot a_{i+1} + 1 = a_{i+2}$ for $i = 1, 2, \ldots, n$.

Problem 3 (By Morteza Saghafian, Iran)
An *anti-Pascal triangle* is an arrangement of numbers in the form of an equilateral triangle in such a way that each number, except those in the last row, is the absolute value of the difference of the two numbers immediately below it. For example, the following arrangement is an anti-Pascal triangle with four rows containing all integers from 1 to 10.

$$
\begin{array}{ccccccc}
 & & & 4 & & & \\
 & & 2 & & 6 & & \\
 & 5 & & 7 & & 1 & \\
8 & & 3 & & 10 & & 9
\end{array}
$$

Determine if there is an anti-Pascal triangle with 2018 rows that contains all the integers from 1 to $1 + 2 + \cdots + 2018$.

Problem 4 (By Gurgen Asatryan, Armenia)
A *spot* is a point (x, y) in the plane such that x, y are both positive integers less than or equal to 20.
At the beginning, each of the 400 spots is empty. Anne and Bob alternately place stones, starting with Anne. On her turn, Anne places a new red stone in an empty spot such that the distance between any two spots occupied by red stones is different from $\sqrt{5}$. On his turn, Bob places a new blue stone in any empty spot. (A spot occupied by a blue stone can be any distance from any other place occupied.) They stop when either of them cannot place a stone.
Find the largest K such that Anne can make sure she places at least K red stones, no matter how Bob places his blue stones.

Problem 5 (By Bayarmagnai Gombodorj, Mongolia)
Let $a_1, a_2, \ldots$ be an infinite sequence of positive integers. Suppose that there exists an integer $N > 1$ such that for each $n \geq N$ the number

$$\frac{a_1}{a_2} + \frac{a_2}{a_3} + \cdots + \frac{a_{n-1}}{a_n} + \frac{a_n}{a_1}$$

is integer. Show that there exists a positive integer M such that $a_m = a_{m+1}$ for all $m \geq M$.

Problem 6 (By Tomasz Ciesla, Poland)
In a convex quadrilateral $ABCD$ the equality $AB \cdot CD = BC \cdot DA$ is hold. The point X inside $ABCD$ is such that $\angle XAB = \angle XCD$ and $\angle XBC = \angle XDA$. Show that $\angle BXA + \angle DXC = 180°$.

IMO 2017

58th International Mathematical Olympiad

Rio de Janeiro, Brazil | July 12 – 23, 2017.

Day 1 (July 18, 2017)

Problem 1 (By Stephan Wagner, South Africa)
For each integer $a_0 > 1$ is defined the sequence $a_0, a_1, a_2, \ldots$ such that for each $n \geq 0$:

$$a_{n+1} = \begin{cases} \sqrt{a_n}, & \text{if } \sqrt{a_n} \text{ is a integer.} \\ a_n + 3, & \text{in any other case.} \end{cases}$$

Determine all the values of a_0 for which there exists a number A such that $a_n = A$ for infinite values of n.

Problem 2 (By Dorlir Ahmeti, Albania)
Let $\mathbb{R}$ be the set of real numbers. Determine all functions $f \colon \mathbb{R} \to \mathbb{R}$ such that, for any real numbers x and y,

$$f(f(x)f(y)) + f(x + y) = f(xy).$$

Problem 3 (By Gerhard Woeginger, Austria)
An invisible rabbit and a hunter play as follows on the Euclidean plane. The rabbit's starting point A_0, and the hunter's starting point B_0 are the same. After $n - 1$ rounds of the game, the rabbit is at point A_{n-1} and the hunter is at point B_{n-1}. In the nth round of the game, three events occur in the following order:
(i) The rabbit moves invisibly to a point A_n such that the distance between A_{n-1} and A_n is exactly 1.

(ii) A tracking device reports a point P_n to the hunter. The only secure information that the device gives the hunter is that the distance between P_n and A_n is less than or equal to 1.

(iii) The hunter visibly moves to a point B_n such that the distance between B_{n-1} and B_n is exactly 1.

Is it always possible that, whatever the way the rabbit moves and whatever points the tracking device reports, the hunter can choose his movements so that after 10^9 rounds the hunter can guarantee that the distance between him and the rabbit is less than or equal to 100?

Day 2 (July 19, 2017)

Problem 4 (By Charles Leytem, Luxembourg)

Let R and S be different points on the circle Ω such that RS is not a diameter of Ω. Let ℓ be the tangent line to Ω at R. Point T is such that S is the midpoint of segment RT. Point J is chosen in the smallest arc RS of Ω so that Γ, the circumscribed circle of triangle JST, intersects ℓ at two different points. Let A be the common point of Γ and ℓ closest to R. Line AJ cuts Ω a second time at K. Show that line KT is tangent to Γ.

Problem 5 (By Grigory Chelnokov, Russia)

Let $N \geq 2$ be a given integer. A group of $N(N + 1)$ soccer players, all of different stature, are lined up. The coach wants to remove $N(N - 1)$ players from this line, so that the resulting line consisting of the remaining $2N$ players satisfies the following N conditions:

(1) No player is between the two tallest players.

(2) No player is between the third tallest player and the fourth tallest player.

...

(N) No player is between the two shorter players.

Prove that this is always possible.

Problem 6 (By John Berman, United States)

An ordered pair (x, y) of integers is defined as a *primitive point* if the greatest common divisor of x and y is 1. Given a finite set S of primitive points, show that there exists a positive integer n and integers $a_0, a_1, \ldots, a_n$, such that for each (x, y) of S, it is true that,

$$a_0 x^n + a_1 x^{n-1} y + a_2 x^{n-2} y^2 + \cdots + a_{n-1} x y^{n-1} + a_n y^n = 1.$$

IMO 2016

57th International Mathematical Olympiad

Hong Kong, Hong Kong | July 06 – 16, 2016.

Day 1 (July 11, 2016)

Problem 1 (By Art Waeterschoot, Belgium)

The triangle BCF is right at B. Let A be a point on the line CF such that $FA = FB$ and F is between A and C. The point D is chosen so that $DA = DC$ and AC is a bisector of the angle $\angle DAB$. Likewise, the point E is chosen so that $EA = ED$ and AD is the bisector of the angle $\angle EAC$. Also, let M be the midpoint of CF and X a point such that $AMXE$ is a parallelogram (with $AM \parallel EX$ and $AE \parallel MX$). Show that the lines BD, FX, and ME are concurrent.

Problem 2 (By Trevor Tao, Australia)

Find all the positive integers n so that in each cell of an $n \times n$ square board one of the letters I, M and O can be written such that:

• in each row and in each column, one third of the cells have I, one third have M and one third have O; and

• in any diagonal line is composed of a number of cells divisible by 3, exactly one third of the cells have I, one third have M and one third have O.

Note: The rows and columns of the n × n square board are numbered 1 through n, in their natural order. Thus, each cell corresponds to a pair of positive integers (i, j) with $1 \leq i, j \leq n$. For n > 1, the board has 4n − 2 diagonal lines of two types. A diagonal line of the first type is composed of all cells (i, j) for which i + j is a constant, while a diagonal line of the second type is composed of all cells (i, j) for which i − j is a constant.

Problem 3 (By Aleksandr Gaifullin, Russia)

Let $P = A_1 A_2 \ldots A_k$ be a convex polygon in the plane. The vertices A_1, A_2, ..., A_k have integer coordinates and lie on a circle. Let S be the area of P. Let n be an odd positive integer such that the squares of the lengths of the sides of P are all integers divisible by n. Show that $2S$ is an integer divisible by n.

Day 2 (July 12, 2016)

Problem 4 (By Gerhard Woeginger, Luxembourg)

A set of positive integers is called *fragrant* if it contains at least two elements, and each of its elements has some prime factor in common with at least one of the remaining elements. Let $P(n) = n^2 + n + 1$. Determine the smallest positive integer b for which there exists some non-negative integer a such that the set $\{P(a+1), P(a+2), \ldots, P(a+b)\}$ is fragrant.

Problem 5 (By Nazar Agakhanov and Ilya Bogdanov, Russia)

On a board is written the next equation,

$$(x-1)(x-2) \cdots (x-2016) = (x-1)(x-2) \cdots (x-2016)$$

which has 2016 linear factors on each side. Determine the smallest possible value of k for which exactly k of these 4032 linear factors can be erased, so that at least one factor remains on each side and the resulting equation has no real solutions.

Problem 6 (By Josef Tkadlec, Czech Republic)

There are $n \geq 2$ segments in the plane such that each pair of segments intersect at a point, and no three segments have a point in common. Jeff has to choose one end of each segment and place a frog on it facing the other end. Then he will whistle $n-1$ times. At each whistle, each frog will immediately jump forward to the next point of intersection on its segment. Frogs never change the directions of their jumps. Jeff wants to position the frogs in such a way that no two of them occupy the same point of intersection at the same time.

(a) Show that if n is odd, Jeff can always achieve his goal.
(b) Show that if n is even, Jeff will never achieve his goal.

IMO 2015

56th International Mathematical Olympiad

Chiang Mai, Thailand | July 04 – 16, 2015.

Day 1 (July 10, 2015)

Problem 1 (By Merlijn Staps, Netherlands)

We say that a finite set of points S in the plane is *balanced* if, for each pair of different points A and B in S there exists a point C in S such that $AC = BC$. Likewise, we say that S is *center-free* if, for each triplet of different points A, B and C in S there is no point P in S such that $PA = PB = PC$.

(a) Show that for all $n \geq 3$ there exists a balanced set of n points.

(b) Determine all integers $n \geq 3$ for which there exists a set of n balanced and center-free points.

Problem 2 (By Dušan Djukić, Serbia)

Find all triples (a, b, c) of positive integers such that each of the numbers $ab - c, \quad bc - a, \quad ca - b$ is a power of 2.

(*A power of 2 is an integer of the form 2^n, where n is a non-negative integer*)

Problem 3 (By Danylo Khilko and Mykhailo Plotnikov, Ukraine)

Let ABC be an acute triangle with $AB > AC$. Let Γ be its circumscribed circle, H its orthocenter, and F the foot of the height from A. Let M be the midpoint of segment BC. Let Q be the point of Γ such that $\angle HQA = 90°$ and let K be the point of Γ such that $\angle HKQ = 90°$. Suppose that points A, B, C, K, and Q are all different and lie on Γ in this order. Show that the circumscribed circle of triangle KQH is tangent to the circumscribed circle of triangle FKM.

Problem 4 (By Silouanos Brazitikos and Evangelos Psychas, Greece)
Triangle ABC has the circumscribed circle Ω with circumcenter O. A circle Γ with center A intersects segment BC at points D and E such that B, D, E, and C are all different and lie on line BC in that order. Let F and G be the intersection points of Γ and Ω, such that A, F, B, C and G lie on Ω in this order. Let K be the second point of intersection of the circumscribed circle of triangle BDF and segment AB. Let L be the second point of intersection of the circumscribed circle of triangle CGE and segment CA. Suppose that lines FK and GL are different and intersect at point X. Show that X lies on line AO.

Problem 5 (By Dorlir Ahmeti, Albania)
Let $\mathbb{R}$ be the set of real numbers. Determine all functions $f : \mathbb{R} \rightarrow \mathbb{R}$ that satisfy the functional equation

$$f(x + f(x + y)) + f(xy) = x + f(x + y) + yf(x)$$

for all real numbers x, y.

Problem 6 (By Ross Atkins and Ivan Guo, Australia)
The sequence of integers $a_1, a_2, \ldots$ satisfies the following conditions:
(i) $1 \leq a_j \leq 2015$ for all $j \geq 1$;
(ii) $k + a_k \neq \ell + a_\ell$ for all $1 \leq k < \ell$.
Show that there are two positive integers b and N such that

$$\left| \sum_{j=m+1}^{n} (a_j - b) \right| \leq 1007^2$$

for all integers m and n that satisfy $n > m \geq N$.

44

IMO 2014

55th International Mathematical Olympiad

Cape Town, South Africa | July 03 – 13, 2014.

Day 1 (July 08, 2014)

Problem 1 (By Gerhard Woeginger, Austria)

Let $a_0 < a_1 < a_2 < \cdots$ be an infinite sequence of positive integers. Prove that there exists a unique integer $n \geq 1$ such that

$$a_n < \frac{a_0 + a_1 + \cdots + a_n}{n} \leq a_{n+1}. \quad (*)$$

Problem 2 (By Tonči Kokan, Croatia)

Let $n \geq 2$ be an integer. Consider a square board of size $n \times n$ made up of n^2 square cells. A configuration of n chips on this board is said to be *peaceful* if there is exactly one chip in each row and each column. Find the largest positive integer k such that, for each peaceful configuration of n chips, there exists a square of size $k \times k$ with no chips in its k^2 cells.

Problem 3 (By Ali Zamani, Iran)

In the convex quadrilateral $ABCD$, we have $\angle ABC = \angle CDA = 90°$. The perpendicular to BD from A intersects BD at point H. Points S and T are on sides AB and AD, respectively; and are such that H is a interior point of triangle SCT and

$$\angle CHS - \angle CSB = 90° \quad , \quad \angle THC - \angle DTC = 90°.$$

Show that line BD is tangent to the circumscribed circle of triangle TSH.

45

Problem 4 (By Giorgi Arabidze, Georgia)

Points P and Q are on side BC of the acute triangle ABC so that $\angle PAB = \angle BCA$ and $\angle CAQ = \angle ABC$. Points M and N are on lines AP and AQ, respectively, so P is the midpoint of AM, and Q is the midpoint of AN. Show that lines BM and CN intersect at the circumscribed circle of triangle ABC.

Problem 5 (By Gerhard Woeginger, Luxembourg)

For each positive integer n, the Bank of Cape Town produces coins of value $1/n$. Given a finite collection of such coins (not necessarily of different values) whose total value does not exceed $99 + 1/2$, show that it is possible to separate this collection into 100 or fewer piles, so that the total value of each pile is at most 1.

Problem 6 (By Gerhard Woeginger, Austria)

A set of lines in the plane is in *general position* if there are not two of them that are parallel or three that pass through the same point. A set of lines in general position separates the plane into regions, some of which have finite area; we call these ones its *finite regions.*

Show that for each n sufficiently large, in any set of n lines in general position, it is possible to color blue at least $\sqrt{n}$ of them in such a way that none of their finite regions has all the sides of their boundary blue.

Note: Solutions that replace $\sqrt{n}$ by $c\sqrt{n}$ will be awarded points depending on the value of c.

IMO 2013

54th International Mathematical Olympiad

Santa Marta, Colombia | July 18 – 28, 2013.

Day 1 (July 23, 2013)

Problem 1 (By Japanese PSC, Japan)

Prove that for any pair of positive integers k and n, there exist k positive integers $m_1, m_2, \ldots, m_k$ (not necessarily different) such that

$$1 + \frac{2^k - 1}{n} = \left(1 + \frac{1}{m_1}\right)\left(1 + \frac{1}{m_2}\right)\cdots\left(1 + \frac{1}{m_k}\right)$$

Problem 2 (By Ivan Guo, Australia)

In a configuration of 4027 points in the plane, where 2013 are red and 2014 are blue, in which there are not three of them that are collinear, it is called *Colombian*. After drawing some lines, the plane is divided into several regions. A collection of lines is *good* if the following conditions are met for a Colombian configuration:
• no line passes through any point of the configuration;
• no region contains points of both colors.
Find the minimum value of k such that for any Colombian configuration of 4027 points, there exists a good collection of k lines.

Problem 3 (By Alexander A. Polyansky, Russia)

Let us assume that the excircle of triangle ABC opposite vertex A is tangent to side BC at point A_1. Similarly, points B_1 in CA and C_1 in AB are defined, considering the excircles opposite to B and C respectively. Likewise, let us assume that the circumcenter of triangle

47

$A_1B_1C_1$ lies on the circle that passes through vertices A, B and C. Prove that triangle ABC is right.

(The excircle of triangle ABC opposite vertex A is the circle that is tangent to side BC, and to the extension of side AB beyond B, and to the extension of side AC beyond C. Similarly, the excircles opposite to the vertices B and C are defined)

Day 2 (July 24, 2013)

Problem 4 (By W. Suksompong and P. Suteparuk, Thailand)

Let ABC be an acute triangle with orthocenter H, and let W be a point on side BC, located between B and C. Points M and N are the feet of the heights drawn from B and C, respectively. The circumscribed circle of triangle BWN is denoted by ω_1, and by X the point of ω_1 such that WX is a diameter of ω_1. Similarly, the circumscribed circle of triangle CWM is denoted by ω_2, and by Y the point of ω_2 such that WY is a diameter of ω_2. Prove that points X, Y and H are collinear.

Problem 5 (By Nikolai Nikolov, Bulgaria)

Let $\mathbb{Q}_{>0}$ be the set of rational numbers greater than zero. And let $f : \mathbb{Q}_{>0} \to \mathbb{R}$ be a function that satisfies the following conditions:

(i) for all $x, y \in \mathbb{Q}_{>0}$, $f(x)f(y) \geq f(xy)$;

(ii) for all $x, y \in \mathbb{Q}_{>0}$, $f(x + y) \geq f(x) + f(y)$;

(iii) there exists a rational number $a > 1$ such that $f(a) = a$.

Prove that $f(x) = x$ for all $x \in \mathbb{Q}_{>0}$.

Problem 6 (By A. S. Golovanov and M. A. Ivanov, Russia)

Let n be an integer such that $n \geq 3$. Let us now consider a circle where $n + 1$ equally spaced points have been marked. Each point is labeled with one of the numbers $0, 1, \ldots, n$ so that each number is used exactly once. Two labeling are considered the same, if one of them can be obtained from the other by a rotation of the circle. A labeling is called *beautiful*, if for any four labels $a < b < c < d$ with $a + d =$

$b + c$, the chord connecting the points labeled a and d does not intersect the chord connecting the points labeled b and c.

Let M be the number of beautiful labelings and N the number of ordered pairs (x, y) of positive integers such that $x + y \leq n$ and $\gcd (x, y) = 1$. Prove that $M = N + 1$.

IMO 2012

53th International Mathematical Olympiad

Mar del Plata, Argentina | July 04 – 16, 2012.

Day 1 (July 10, 2012)

Problem 1 (By Evangelos Psychas, Greece)
Given a triangle ABC, point J is the center of the excircle opposite vertex A. This excircle is tangent to side BC at M, and to lines AB and AC at K and L, respectively. Lines LM and BJ intersect at F, and lines KM and CJ intersect at G. Let S be the point of intersection of lines AF and BC, and let T be the point of intersection of lines AG and BC. Show that M is the midpoint of ST.
(The excircle of ABC opposite vertex A is the circle that is tangent to side BC, to the extension of side AB beyond B, and to the extension of side AC beyond C.)

Problem 2 (By Angelo di Pasquale, Australia)
Let $n \geq 3$ be an integer, and let $a_2, a_3, \ldots, a_n$ positive real numbers such that $a_2 \cdot a_3 \cdot \ldots \cdot a_n = 1$. Show that

$$(1 + a_2)^2 (1 + a_3)^3 \cdots (1 + a_n)^n > n^n.$$

Problem 3 (By David Arthur, Canada)
The liar's guessing game is a game played between two players A and B. The rules of the game depend on two positive integers k and n that are known to both players. At the beginning of the game, player A chooses integers x and N such that $1 \leq x \leq N$. Player A keeps x secret, and reveals to B the real value of N. Next, player B tries to obtain information about x by asking A questions as follows: in each question, B specifies an arbitrary set S of positive integers (which can

51

be one of those specified in a previous question), and asks A if x belongs to S. Player B can ask as many such questions as he wants. After each question, player A must immediately answer it with yes or no, but he can lie as many times as he wants. The only restriction is that among any $k+1$ consecutive answers, at least one must be true. After B has asked as many questions as he wishes, he must specify a set X of at most n positive integers. If x belongs to X then B wins; otherwise, he loses. Show that,

(1) If $n \geq 2^k$, then B can be guaranteed a victory, and

(2) For every sufficiently large k, there exists an integer $n \geq 1.99^k$ such that B cannot be guaranteed a victory.

Day 2 (July 11, 2012)

Problem 4 (By Liam Baker, South Africa)

Find all the functions $f: \mathbb{Z} \to \mathbb{Z}$ that satisfy the following equality:

$$f(a)^2 + f(b)^2 + f(c)^2 = 2f(a)f(b) + 2f(b)f(c) + 2f(c)f(a),$$

for all integers a, b, c where $a + b + c = 0$. ($\mathbb{Z}$ denotes the set of whole numbers.)

Problem 5 (By Josef Tkadlec, Czech Republic)

Let ABC be a triangle such that $\angle BCA = 90°$, and let D be the foot of the height from C. Let X be an interior point of segment CD. Let K be a point on segment AX such that $BK = BC$. Similarly, let L be a point on segment BX such that $AL = AC$. Let M be the point of intersection of AL and BK. Prove that $MK = ML$.

Problem 6 (By Dušan Djukić, Serbia)

Find all positive integers n for which there are non-negative integers $a_1, a_2, \ldots, a_n$ such that

$$\frac{1}{2^{a_1}} + \frac{1}{2^{a_2}} + \cdots + \frac{1}{2^{a_n}} = \frac{1}{3^{a_1}} + \frac{2}{3^{a_2}} + \cdots + \frac{n}{3^{a_n}} = 1.$$

IMO 2011

52th International Mathematical Olympiad

Amsterdam, Netherlands | July 12 – 24, 2011.

Day 1 (July 18, 2011)

Problem 1 (By Fernando Campos, Mexico)
Given a set $A = \{a_1, a_2, a_3, a_4\}$ of four different positive integers, the sum $a_1 + a_2 + a_3 + a_4$ is denoted by S_A. Let n_A be the number of pairs (i, j) such that $1 \leq i < j \leq 4$, so that $a_i + a_j$ divides S_A. Find all sets A of four different positive integers so that the largest possible value of n_A is reached.

Problem 2 (By Geoff Smith, United Kingdom)
Let S be a finite set of at least two points in the plane. Also, in S there are no three collinear points. A *whirlpool* is a process that begins with a line ℓ that passes through a single point P of S. Line ℓ is rotated clockwise with center at P until the line encounters another point of S for the first time, which we will call Q. With Q as a new center, keep rotating the line clockwise until the line finds another point on S. This process continues indefinitely. Show that you can choose a point P in S and a line ℓ through P such that the resulting whirlpool uses each point in S as a center of rotation an infinite number of times.

Problem 3 (By Igor Voronovich, Belarus)
Let $f: \mathbb{R} \longrightarrow \mathbb{R}$ be a function defined on the set of real numbers, which satisfies the expression $f(x + y) \leq y f(x) + f(f(x))$, for every pair of real numbers x, y. Prove that $f(x) = 0, \forall x \leq 0$.

Problem 4 (By Morteza Saghafiyan, Iran)

Let n be an integer such that $n > 0$. There is a two-plate scale and n weights whose weights are $2^0, 2^1, \ldots, 2^{n-1}$. We must place each of the n weights on the scale, one after another, in such a way that the right plate is never heavier than the left plate. At each step, we choose one of the weights that has not been placed on the scale, and we place it on either the left plate or the right plate, until all the weights have been placed. Find out the number of ways this can be done.

Problem 5 (By Seyedmahyar Sefidgaran, Iran)

Let $f: \mathbb{Z} \longrightarrow \mathbb{Z}^+$, a function from the set of integers to the set of positive integers. Likewise, for any two integers m and n, the difference $f(m) - f(n)$ is divisible by $f(m - n)$.

Prove that for all integers m and n with $f(m) \le f(n)$, the number $f(n)$ is divisible by $f(m)$.

Problem 6 (By Japanese PSC, Japan)

Let ABC be an acute triangle whose circumscribed circle is Γ. Let ℓ be a tangent line to Γ, and let ℓ_a, ℓ_b and ℓ_c be the lines which are obtained by reflection of ℓ with respect to the lines BC, CA and AB, respectively. Show that the circumscribed circle of the triangle determined by the lines ℓ_a, ℓ_b and ℓ_c is tangent to the circle Γ.

IMO 2010

51th International Mathematical Olympiad

Astana, Kazakhstan | July 02 – 14, 2010.

Day 1 (July 07, 2010)

Problem 1 (By Pierre Bornsztein, France)

Determine all functions $f: \mathbb{R} \rightarrow \mathbb{R}$ such that the equality

$$f(\lfloor x \rfloor y) = f(x) \lfloor f(y) \rfloor$$

is satisfied for all numbers $x, y \in \mathbb{R}$. ($\lfloor z \rfloor$ denotes the largest integer that is less than or equal to z.)

Problem 2 (By Tai Wai Ming and Wang Chongli, Hong Kong)

Given a triangle ABC, let I be its incenter and Γ its circumscribed circle. Line AI intersects Γ again at D. Let E be a point on the arc $\overset{\frown}{BDC}$ and F a point on side BC such that

$$\angle BAF = \angle CAE < \frac{1}{2} \angle BAC.$$

Further, G is the midpoint of the segment IF. Show that lines DG and EI intersect at Γ.

Problem 3 (By Gabriel Carroll, United States)

Let $\mathbb{N}$ be the set of positive integers. Determine all functions $g: \mathbb{N} \rightarrow \mathbb{N}$ such that $(g(m) + n)(m + g(n))$ is a perfect square for all $m, n \in \mathbb{N}$.

Problem 4 (By Marcin Kuczma, Poland)

Given the triangle ABC, let Γ be its circumscribed circle and P be an interior point. Lines AP, BP and CP intersect Γ again at points K, L, and M, respectively. The tangent line to Γ at C intersects the line AB at S. If $SC = SP$ is satisfied, prove that $MK = ML$.

Problem 5 (By Hans Zantema, Netherlands)

In each of six boxes $B_1, B_2, B_3, B_4, B_5, B_6$ there is initially a single coin. Two types of operations are allowed:

Type 1: Choose a non-empty box B_j, with $1 \leq j \leq 5$. Remove a coin from B_j and add two coins to B_{j+1}.

Type 2: Choose a non-empty box B_k, with $1 \leq k \leq 4$. Remove a coin from B_k and exchange the contents of the (possibly empty) boxes B_{k+1} and B_{k+2}.

Determine if there is a finite sequence of these operations that leaves boxes B_1, B_2, B_3, B_4, B_5 empty and box B_6 containing exactly $2010^{2010^{2010}}$ coins. (Note that $a^{b^c} = a^{(b^c)}$.)

Problem 6 (By Morteza Saghafiyan, Iran)

Let $a_1, a_2, a_3, \ldots$ be a sequence of positive real numbers. For a certain positive integer s,

$$a_n = max\{a_k + a_{n-k} \text{ such that } 1 \leq k \leq n - 1\}$$

for all $n > s$. Prove that there exist positive integers ℓ and N, with $\ell \leq s$, such that $a_n = a_\ell + a_{n-\ell}$ for all $n \geq N$.

En Español

Problemas
IMO 2020

IMO 2020

61° Olimpiada Internacional de Matemáticas

San Petersburgo, Rusia | 19 – 28 de Septiembre, 2020.

Día 1 (21 de Sept., 2020)

Problema 1 (Por Dominik Burek, Polonia)

En el interior de un cuadrilátero convexo $ABCD$, hay un punto P tal que se cumplen las siguientes igualdades:

$$\angle PAD : \angle PBA : \angle DPA = 1 : 2 : 3 = \angle CBP : \angle BAP : \angle BPC.$$

Demuestre que las siguientes tres líneas se intersecan en un punto: las bisectrices internas de los ángulos ADP y PCB y la mediatriz del segmento AB.

Problema 2 (Por Stijn Cambie, Bélgica)

Dados los números reales a, b, c, d tales que $a \geq b \geq c \geq d > 0$ y $a + b + c + d = 1$. Demostrar que,

$$(a + 2b + 3c + 4d)\, a^a \cdot b^b \cdot c^c \cdot d^d < 1.$$

Problema 3 (Por Carl Schildkraut and Milan Haiman, Hungría)

Hay $4n$ guijarros con masas $1, 2, 3, \ldots, 4n$. Cada uno de los guijarros está pintado en uno de n colors, y hay 4 guijarros de cada color. Demostrar que los guijarros se pueden dividir en dos montones de igual peso total de modo que cada montón contenga dos guijarros de cada color.

Problema 4 (Por Tejaswi Navrinarekallu, India)

Dado un número entero $n > 1$. Hay n^2 estaciones de teleféricos en la ladera de una montaña a diferentes altitudes. Cada una de las dos compañías de teleféricos A y B posee k teleféricos. Cada teleférico realiza un traslado regular sin escalas de una estación a otra superior. Se sabe que k traslados de la compañía A comienzan en k estaciones diferentes y terminan en k estaciones diferentes; además, los traslados que comienzan en la parte superior terminan también en la parte superior. Se cumplen las mismas condiciones para la compañía B. Diremos que dos estaciones están *conectadas* por una compañía de teleféricos si es posible ir de la estación inferior a la superior utilizando uno o más traslados de esta compañía (otros traslados entre estaciones están prohibidos). Determine el menor entero k de modo que haya dos estaciones conectadas por ambas compañías.

Problema 5 (Por Oleg Košik, Estonia)

Hay $n > 1$ cartas en una baraja, cada una de las cuales contiene un número entero positivo. Resultó que, para dos cartas cualesquiera, la media aritmética de los números escritos en ellas es igual a la media geométrica de los números escritos en las cartas de un conjunto dado que consta de una o más cartas. ¿Para qué n son iguales todos los números escritos en las cartas?

Problema 6 (Por Ting-Feng Lin and Hung-Hsun Hans Yu, Taiwan)

Demuestre que hay una constante positiva c para la que se cumple el siguiente enunciado:
Sea S un conjunto de $n > 1$ puntos del plano, en el que la distancia entre dos puntos cualesquiera no es menor que 1. Entonces hay una recta ℓ que separa el conjunto S tal que la distancia desde cualquier punto de S a ℓ no es menor que $cn^{-1/3}$.

(Una recta ℓ *separa* el conjunto de puntos S si corta algún segmento cuyos extremos pertenecen a S.)

Nota. A los resultados más débiles obtenidos al reemplazar $cn^{-1/3}$ por $cn^{-1/\alpha}$ se les podrá otorgar puntos dependiendo del valor de la constante $\alpha > 1/3$.

Solución
IMO 2020

Problema 1 (Por Dominik Burek, Polonia)

En el interior de un cuadrilátero convexo ABCD, hay un punto P tal que se cumplen las siguientes igualdades:

$$\angle PAD : \angle PBA : \angle DPA = 1 : 2 : 3 = \angle CBP : \angle BAP : \angle BPC.$$

Demuestre que las siguientes tres líneas se intersecan en un punto: las bisectrices internas de los ángulos ADP y PCB y la mediatriz del segmento AB.

Solución

Sea $\varphi = \angle PAD$ y $\psi = \angle CBP$; luego $\angle PBA = 2\varphi$, $\angle DPA = 3\varphi$, $BAP = 2\psi$ y $\angle BPC = 3\psi$ (ver Fig. 1). Sea X un punto del segmento AD tal que $\angle XPA = \varphi$. Tenemos

$$\angle PXD = \angle PAX + \angle XPA = 2\varphi = \angle DPA - \angle XPA = \angle DPX.$$

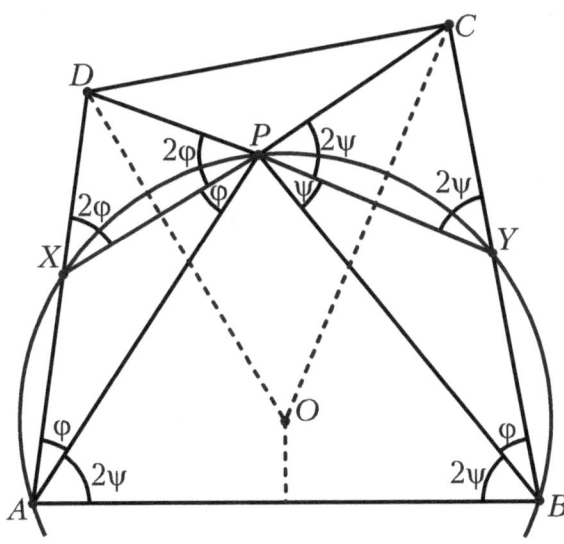

Fig. 1

Entonces el triángulo DPX es isósceles con lados $DX = DP$; por lo tanto, la bisectriz del ángulo ADP es la mediatriz de XP. De manera similar, si Y es un punto en BC tal que $\angle BPY = \psi$, entonces la bisectriz del ángulo PCB es la mediatriz de PY. Luego, necesitamos probar que las mediatrices de los segmentos XP, PY y AB se cortan en un punto.

Asimismo, se observa que $\angle AXP = 180° - \angle PXD = 180° - 2\varphi = 180° - \angle PBA$. Por lo tanto, el cuadrilátero $AXPB$ es inscriptible, y X está en la circunferencia circunscrita del triángulo APB. De manera similar, Y se encuentra en la circunferencia circunscrita del triángulo APB. Por lo tanto, las mediatrices de los segmentos XP, PY, y AB pasan por el centro de la circunferencia ($ABYPX$). Completándose así la demostración.

Problema 2 (Por Stijn Cambie, Bélgica)

Dados los números reales a, b, c, d tales que $a \geq b \geq c \geq d > 0$ y $a + b + c + d = 1$. Demostrar que,

$$(a + 2b + 3c + 4d)\, a^a \cdot b^b \cdot c^c \cdot d^d < 1.$$

Solución

Aplicamos la desigualdad de medias (ponderada) generalizada a los reales positivos $x_1, \ldots, x_n$ y $\lambda_1, \ldots, \lambda_n$ tal que $\lambda_1 + \cdots + \lambda_n = 1$,

$$x_1^{\lambda_1} \cdot x_2^{\lambda_2} \cdot \ldots \cdot x_n^{\lambda_n} \leq \lambda_1 x_1 + \lambda_2 x_2 + \cdots + \lambda_n x_n.$$

Tenemos que

$$a^a \cdot b^b \cdot c^c \cdot d^d \leq a \cdot a + b \cdot b + c \cdot c + d \cdot d = a^2 + b^2 + c^2 + d^2,$$

por tanto, basta probar que

$$(a + 2b + 3c + 4d)\,(a^2 + b^2 + c^2 + d^2) < 1 = (a + b + c + d)^3.$$

La siguiente desigualdad

$$(a + b + c + d)^3 \geq (a + 2b + 3c + 4d)\,(a^2 + b^2 + c^2 + d^2)$$

se puede probar, por ejemplo, como sigue. Tenemos

$$(a + b + c + d)^3 > a^2(a + 3b + 3c + 3d) + b^2(3a + b + 3c + 3d) +$$
$$+ c^2(3a + 3b + c + 3d) + d^2(3a + 3b + 3c + d),\ (*)$$

ya que todos los términos del lado derecho están explícitamente presentes después de expandir los corchetes del lado izquierdo. Dado que cada una de las expresiones $a + 3b + 3c + 3d$, $3a + b + 3c + 3d$, $3a + 3b + c + 3d$, $3a + 3b + 3c + d$ no es menor que $a + 2b + 3c + 4d$; por consiguiente, el lado derecho $(*)$ no es menor que $(a + 2b + 3c + 4d)\,(a^2 + b^2 + +d^2)$.

Problema 3 (Por Carl Schildkraut y Milan Haiman, Hungría)

Hay $4n$ guijarros con masas $1, 2, 3, \ldots, 4n$. Cada uno de los guijarros está pintado en uno de n colors, y hay 4 guijarros de cada color. Demostrar que los guijarros se pueden dividir en dos montones de igual peso total de modo que cada montón contenga dos guijarros de cada color.

Solución

Juntamos los guijarros en pares con suma de masas $4n + 1$, de modo que obtenemos un conjunto S que consta de $2n$ pares: $(1, 4n)$, $(2, 4n - 1)$, $(3, 4n - 2)$, ..., $(2n, 2n + 1)$. Basta con dividir S en dos subconjuntos de n pares de manera que cada subconjunto contenga dos guijarros de cada color. Introduzcamos un multigrafo G (es decir, un grafo en el que se permiten bucles y aristas múltiples) en n vertices vértices de modo que cada vértice corresponda a un determinado color. Para cada par de guijarros del conjunto S, trazamos una arista entre los vértices correspondientes a los colores de estos guijarros. En este caso, el grado de cada vértice será 4. Además, tengamos en cuenta que la división requerida de los guijarros corresponde a colorear las aristas del grafo en dos colores, por ejemplo, azul y rojo, de modo que de cada vértice surgen exactamente 2 aristas rojas y 2 azules.

Para completar la solución, es suficiente hacer coincidir un color con cada componente conectado G^c del grafo G. Dado que los grados de todos los vértices son pares, hay una ruta de Euler P en G^c (es decir, un ciclo que pasa a través de cada arista de G^c exactamente una vez). Notemos que el número de aristas en P es par (es igual al doble del número de vértices en G^c). Por lo tanto, todas las aristas se pueden colorear de rojo y azul de modo que dos aristas adyacentes de P tengan colores diferentes (se puede recorrer a lo largo de P y colorear las aristas, alternando rojo y azul). Por lo tanto, el mismo número de aristas rojas y azules surgen de cada vértice en G^c, como se desea.

Problema 4 (Por Tejaswi Navrinarekallu, India)

Dado un número entero $n > 1$. Hay n^2 estaciones de teleféricos en la ladera de una montaña a diferentes altitudes. Cada una de las dos compañías de teleféricos A y B posee k teleféricos. Cada teleférico realiza un traslado regular sin escalas de una estación a otra superior. Se sabe que k traslados de la compañía A comienzan en k estaciones diferentes y terminan en k estaciones diferentes; además, los traslados que comienzan en la parte superior terminan también en la parte superior. Se cumplen las mismas condiciones para la compañía B. Diremos que dos estaciones están "conectadas" por una compañía de teleféricos si es posible ir de la estación inferior a la superior utilizando uno o más traslados de esta compañía (otros traslados entre estaciones están prohibidos). Determine el menor entero k de modo que haya dos estaciones conectadas por ambas compañías.

Solución

Respuesta. $k = n^2 - n - 1$.

En primer lugar, probaremos que para cualquier $k \leq n^2 - n$, puede que no haya un par de estaciones conectadas por cada una de las dos compañías. Basta con dar un ejemplo para $k = n^2 - n$.

Suponga que la compañía A tiene traslados $(i, \ i + 1)$, donde i no es divisible por n, y que la compañía B tiene traslados de la forma $(i, \ i + n)$, donde $1 \leq i \leq n^2 - n$. Luego, los números i, j de las estaciones conectadas por la compañía A son tales que $|i - j| \leq n - 1$. Está claro que estas estaciones no están conectadas por la compañía B.

Tengamos en cuenta que todas las estaciones finales de los traslados A son diferentes, por lo tanto, hay $n^2 - k = n - 1$ estaciones $a_1, \ldots, a_{n-1}$, que no son finales para los traslados A. Desde la estación a_1 subiremos con traslados A lo más alto posible. Esta ruta da la máxima cadena A de estaciones. De manera similar, cada una de las estaciones $a_2, \ldots, a_{n-1}$ determina la máxima cadena A (posiblemente consistente de una sola estación). Está claro que cada una de las n^2

71

estaciones están incluidas en una de las cadenas máximas A (ya que para cada estación el recorrido de bajada por traslados A a una de las estaciones $a_1, \ldots, a_{n-1}$ está únicamente determinado). Análogamente, podemos probar que cada estación pertenece a una de las $n-1$ cadenas máximas B. Notemos que una de las cadenas máximas A contiene al menos $n^2 / (n-1) > n$ estaciones, y al menos dos de estas estaciones pertenecen a una misma cadena máxima B. Finalmente, las dos estaciones encontradas son las requeridas.

Problema 5 (Por Oleg Košik, Estonia)

Hay $n > 1$ cartas en una baraja, cada una de las cuales contiene un número entero positivo. Resultó que, para dos cartas cualesquiera, la media aritmética de los números escritos en ellas es igual a la media geométrica de los números escritos en las cartas de un conjunto dado que consta de una o más cartas. ¿Para qué n son iguales todos los números escritos en las cartas?

Solución

Respuesta. Para todo n.

Suponga que no todos los números son iguales, y que $b_1 > b_2 > \ldots > b_k$ son todos números diferentes entre los números de las cartas, $k > 2$ (de modo que el conjunto de números de las cartas consta de varias copias de los números b_1, varias copias de números b_2, etc.). Si $d = mcd\,(b_1, \ldots, b_k) > 1$, dividiendo todos los números en las cartas por d, llegamos a un ejemplo que satisface la condición y en el que el *mcd* de todos los números de las cartas es 1. Luego, en lo que sigue supondremos que $mcd\,(b_1, \ldots, b_k) = 1$. Sea p un divisor primo de b_1 ((existe, ya que $b_1 > 1$). Dado que $gcd\,(b_1, \ldots, b_k) = 1$, hay un mínimo $m \geq 2$ tal que b_m no es un múltiplo de p.

El número $q = (b_{m-1} + b_m)/2$ tiene que ser la media geométrica de varios b_i, es decir, $q^t = c_1 c_2 \ldots c_t$ para cierto $t \in \mathbb{N}$ y $c_i \in \{b_1, \ldots, b_k\}$. Tengamos en cuenta que algunos c_i coinciden con uno de los números $b_1, \ldots, b_{m-1}$, porque de lo contrario $c_1 c_2 \ldots c_t \leq b_m^t < q^t$. Por lo tanto, p es un divisor de ciertos c_i, luego, $c_1 c_2 \ldots c_t$ es divisible por p. Asimismo, la expresión $q^t = c_1 c_2 \ldots c_t$ se reduce a la siguiente igualdad en enteros:

$$(b_{m-1} + b_m)^t = 2^t \cdot c_1 c_2 \ldots c_t.$$

Puesto que el lado derecho es divisible por p, luego $b_{m-1} + b_m$ es divisible por p. Pero por otro lado, por la definición de m, b_{m-1} es divisible por p, pero b_m no lo es. Tenemos una contradicción; entonces es imposible que para cualquier $n > 1$ haya números distintos en las cartas que satisfagan la condición. Finalmente, para todo n, los números en las cartas deben ser iguales.

Problema 6 (Por Ting-Feng Lin and Hung-Hsun Hans Yu, Taiwan)
Demuestre que hay una constante positiva c para la que se cumple el siguiente enunciado:
Sea S un conjunto de $n > 1$ puntos del plano, en el que la distancia entre dos puntos cualesquiera no es menor que 1. Entonces hay una recta ℓ que separa el conjunto S tal que la distancia desde cualquier punto de S a ℓ no es menor que $cn^{-1/3}$.
(Una recta ℓ "separa" el conjunto de puntos S si corta algún segmento cuyos extremos pertenecen a S.)
Nota. A los resultados más débiles obtenidos al reemplazar $cn^{-1/3}$ por $cn^{-1/\alpha}$ se les podrá otorgar puntos dependiendo del valor de la constante $\alpha > 1/3$.

Solución
Probaremos el enunciado requerido para $c = 1/8$. Hagamos $\delta = 1/8 \cdot n^{-1/3}$. Para cualquier recta ℓ y cualquier punto X, denotamos con X_ℓ a la proyección de X sobre ℓ; usamos una notación similar para conjuntos de puntos.

Supongamos que para cierta recta ℓ el conjunto S_ℓ contiene dos puntos vecinos X y Y con una distancia $XY = 2d$. Luego, la mediatriz del segmento XY separa S, y todos los puntos del conjunto S estarán a una distancia de al menos d de la línea recta ℓ. Por lo tanto, si $d \geq \delta$, entonces se ha encontrado la línea requerida. Supongamos, por el contrario, que para cualquier recta ℓ del conjunto S_ℓ, la distancia entre puntos adyacentes es menor que 2δ.

Elegimos los puntos $A, B \in S$ con una distancia máxima entre sí $M = AB$, de modo que AB es el diámetro del conjunto S (ver Fig. 2); obviamente, $M \geq 1$. Sea ℓ la recta AB. El conjunto S está en la intersección de dos círculos D_A y D_B de radio M centrados en los puntos A y B. Por lo tanto, la proyección S_ℓ se encuentra en el segmento AB. Además, los puntos del conjunto S_ℓ dividen el segmento AB en a lo sumo $n - 1$ partes, la longitud de cada una de las cuales es menor que 2δ. Así tenemos

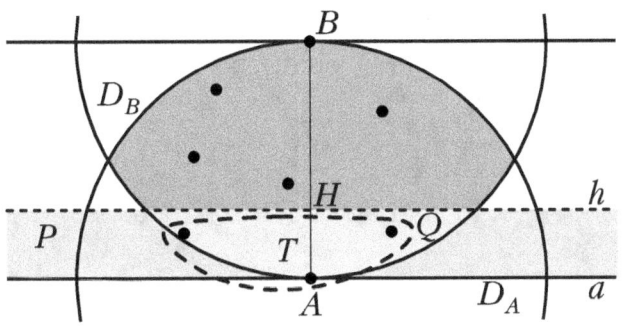

Fig. 2

$$M < n \cdot 2\delta. \tag{1}$$

En el segmento AB, elegimos un punto H tal que $AH = 1/2$. Sea P una franja entre las líneas a y h que pasan por A y H perpendiculares a AB (asumimos que los límites de la franja P le pertenecen). Sea $T = P \cap S$ y $t = |T|$. De acuerdo con nuestro supuesto, el segmento AH no contiene menos de $\lceil 1/2 : (2\delta) \rceil$ puntos del conjunto S_ℓ, de donde

$$t \geq \frac{1}{4\delta}. \tag{2}$$

Nótese que T está contenido en $Q = P \cap D_B$. El conjunto Q es un segmento y su proyección Q_a es un segmento de longitud

$$2\sqrt{M^2 - \left(M - \frac{1}{2}\right)^2} < 2\sqrt{M}.$$

Por otro lado, para dos puntos cualesquiera $X, Y \in T$, tenemos $XY \geq 1$ y $X_\ell Y_\ell \leq 1/2$, lo que significa que $X_a Y_a = \sqrt{XY^2 - X_\ell Y_\ell^2} \geq \sqrt{3}/2$. Por tanto, los t puntos que forman el conjunto T_a están en un intervalo de longitud menor que $2\sqrt{M}$, y están separados entre sí por al menos $\sqrt{3}/2$. Por lo tanto, $2\sqrt{M} > (t - 1)\sqrt{3}/2$, o

$$t < 1 + \frac{4\sqrt{M}}{\sqrt{3}} < 4\sqrt{M}, \qquad\qquad (3)$$

Puesto que $M \geq 1$.

De las expresiones (1), (2) y (3), obtenemos

$$\frac{1}{4\delta} \leq t < 4\sqrt{M} < 4\sqrt{2n\delta},$$

por tanto $512n\delta^3 > 1$, lo cual no es cierto para δ.

Apéndice:

Enunciados
IMO 2010 – 2019

IMO 2019

60° Olimpiada Internacional de Matemáticas

Bath, Reino Unido | 11 – 22 de Julio, 2019.

Día 1 (16 de Julio, 2019)

Problema 1 (Por Liam Baker, Sudáfrica)

Sea $\mathbb{Z}$ el conjunto de números enteros. Hallar todas las funciones $f\colon \mathbb{Z} \longrightarrow \mathbb{Z}$ tal que, para todos los enteros x y y,

$$f(2x) + 2f(y) = f(f(x + y)).$$

Problema 2 (Por Anton Trygub, Ucrania)

En el triángulo ABC, el punto A_1 está en el lado BC y el punto B_1 está en el lado AC. Sean P y Q puntos en los segmentos AA_1 y BB_1, respectivamente, tal que PQ es paralelo a AB. Y sea P_1 un punto en la recta PB_1 diferente a B_1, con B_1 entre P y P_1, y $\angle PP_1C = \angle BAC$. Similarmente, sea Q_1 un punto del segmento QA_1 diferente de A_1, con A_1 entre Q y Q_1, y $\angle CQ_1Q = \angle CBA$. Probar que los puntos P, Q, P_1 y Q_1 son concíclicos.

Problema 3 (Por Adrian Beker, Croacia)

Una red social tiene 2019 usuarios, algunos de los cuales son amigos. Siempre que el usuario A es amigo del usuario B, el usuario B es también amigo del usuario A. Asimismo, pueden ocurrir eventos del siguiente tipo en forma repetida, pero uno a la vez:

Tres usuarios A, B y C de modo que A es amigo de B y C, pero B y C no son amigos, cambian su relación de amistad de manera que B y C ahora son amigos, pero A ya no es amigo ni de B ni de C. Las otras relaciones de amistad no cambian.

79

Inicialmente hay 1010 usuarios que tienen 1009 amigos cada uno, y hay 1009 usuarios que tienen 1010 amigos cada uno. Demostrar que existe una sucesión de este tipo de eventos después de la cual cada usuario es amigo a lo sumo de uno de los demás usuarios.

Dia 2 (17 de Julio, 2019)

Problema 4 (Por Gabriel Chicas Reyes, El Salvador)

Hallar todos los pares de enteros positivos (k, n) que satisfacen la igualdad

$$k! = (2^n - 1)(2^n - 2)(2^n - 4)...(2^n - 2^{n-1}). \qquad (*)$$

Problema 5 (por David Altizio, Estados Unidos)

El Banco de Bath emite monedas con una H en una cara y una T en la otra. Harry tiene n monedas de este tipo, alineadas de izquierda a derecha. Asimismo, él realiza repetidamente la siguiente operación: si hay exactamente k ($k > 0$) monedas que muestren una H, Harry voltea la k-ésima moneda contando desde la izquierda; de lo contrario todas las monedas muestran una T y se detiene el proceso. Por ejemplo, si $n = 3$ y la configuración inicial es THT, el proceso sería $THT \rightarrow HHT \rightarrow HTT \rightarrow TTT$, deteniéndose luego de tres operaciones.

(a) Demostrar que para cualquier configuración inicial que tenga Harry, el proceso se detiene luego de un número finito de operaciones.

(b) Para cada configuración inicial K, sea $L(K)$ el número de operaciones que se realizan hasta que el proceso se detiene. Por ejemplo, $L(THT) = 3$ y $L(TTT) = 0$. Hallar el valor promedio de $L(K)$ de todas las 2^n posibles configuraciones iniciales de K.

Problema 6 (Por Anant Mudgal, India)

Sea I el incentro del triángulo acutángulo ABC tal que $AB \neq AC$. Y sea ω la circunferencia inscrita del triángulo ABC, la cual es tangente a los lados BC, CA y AB en los puntos D, E y F, respectivamente. Asimismo, la recta que pasa por D y perpendicular a EF, interseca a ω nuevamente en R. Además, el segmento AR interseca otra vez a ω en P. Y las circunferencias circunscritas de los triángulos PCE y PBF se intersecan de nuevo en Q. Demostrar que las líneas DI y PQ se cortan en la recta que pasa por A y es perpendicular a AI.

IMO 2018

59° Olimpiada Internacional de Matemáticas

Cluj-Napoca, Rumania | 03 – 14 de Julio, 2018.

Día 1 (09 de Julio, 2018)

Problema 1 (Por S. Brazitikos, E. Psychas y M. Sarantis, Grecia)
Sea Γ la circunferencia circunscrita al triángulo acutángulo ABC. Los puntos D y E están en los segmentos AB y AC, respectivamente, y son tales que $AD = AE$. Las mediatrices de BD y CE cortan a los arcos menores AB y AC de Γ en los puntos F y G, respectivamente. Demostrar que las rectas DE y FG son paralelas (o son la misma recta).

Problema 2 (Por Patrik Bak, Eslovaquia)
Hallar todos los enteros $n \geq 3$ para los que existen números reales $a_1, a_2, \ldots, a_{n+2}$, tales que $a_{n+1} = a_1$ y $a_{n+2} = a_2$, si $a_i \cdot a_{i+1} + 1 = a_{i+2}$ para $i = 1, 2, \ldots, n$.

Problema 3 (Por Morteza Saghafian, Irán)
Un *triángulo anti-Pascal* es una disposición de números en forma de triángulo equilátero de tal manera que cada número, excepto los de la última fila, es el valor absoluto de la diferencia de los dos números que están inmediatamente debajo de este. Por ejemplo, la siguiente disposición es un triángulo anti-Pascal con cuatro filas que contiene todos los enteros desde 1 hasta 10.

$$4$$
$$2 \quad 6$$
$$5 \quad 7 \quad 1$$
$$8 \quad 3 \quad 10 \quad 9$$

Determinar si existe un triángulo anti-Pascal con 2018 filas que contenga todos los enteros desde 1 hasta $1 + 2 + ... + 2018$.

Día 2 (10 de Julio, 2018)

Problema 4 (Por Gurgen Asatryan, Armenia)

Un *lugar* es un punto (x, y) en el plano tal que x, y son ambos enteros positivos menores o iguales que 20.

Al comienzo, cada uno de los 400 lugares está vacío. Ana y Beto colocan piedras alternadamente, comenzando con Ana. En su turno, Ana coloca una nueva piedra roja en un lugar vacío tal que la distancia entre cualesquiera dos lugares ocupados por piedras rojas es distinto de $\sqrt{5}$. En su turno, Beto coloca una nueva piedra azul en cualquier lugar vacío. (Un lugar ocupado por una piedra azul puede estar a cualquier distancia de cualquier otro lugar ocupado.) Ellos paran cuando alguno de los dos no pueda colocar una piedra.

Hallar el mayor K tal que Ana pueda asegurarse de colocar al menos K piedras rojas, sin importar cómo Beto coloque sus piedras azules.

Problema 5 (Por Bayarmagnai Gombodorj, Mongolia)

Sea $a_1, a_2, ...$ una sucesión infinita de enteros positivos. Supongamos que existe un entero $N > 1$ tal que para cada $n \geq N$ el número

$$\frac{a_1}{a_2} + \frac{a_2}{a_3} + \cdots + \frac{a_{n-1}}{a_n} + \frac{a_n}{a_1}$$

es entero. Demostrar que existe un entero positivo M tal que $a_m = a_{m+1}$ para todo $m \geq M$.

Problema 6 (Por Tomasz Ciesla, Polonia)

En un cuadrilátero convexo $ABCD$ se cumple que $AB \cdot CD = BC \cdot DA$. El punto X en el interior de $ABCD$ es tal que

$$\angle XAB = \angle XCD \qquad y \qquad \angle XBC = \angle XDA.$$

Demostrar que $\angle BXA + \angle DXC = 180°$.

IMO 2017

58° Olimpiada Internacional de Matemáticas

Rio de Janeiro, Brasil | 12 – 23 de Julio, 2017.

Día 1 (18 de Julio, 2017)

Problema 1 (Por Stephan Wagner, Sudáfrica)

Para cada entero $a_0 > 1$, se define la sucesión $a_0, a_1, a_2, \ldots$ tal que para cada $n \geq 0$:

$$a_{n+1} = \begin{cases} \sqrt{a_n}, & \text{si } \sqrt{a_n} \text{ es entero.} \\ a_n + 3, & \text{cualquier otro caso.} \end{cases}$$

Determinar todos los valores de a_0 para los que existe un número A tal que $a_n = A$ para infinitos valores de n.

Problema 2 (Por Dorlir Ahmeti, Albania)

Sea $\mathbb{R}$ el conjunto de los números reales. Determinar todas las funciones $f : \mathbb{R} \to \mathbb{R}$ tales que, para cualesquiera reales x y y,

$$f(f(x)f(y)) + f(x + y) = f(xy).$$

Problema 3 (Por Gerhard Woeginger, Austria)

Un conejo invisible y un cazador juegan como sigue en el plano euclídeo. El punto de partida A_0 del conejo, y el punto de partida B_0 del cazador son el mismo. Después de $n - 1$ rondas del juego, el conejo se encuentra en el punto A_{n-1} y el cazador se encuentra en el punto B_{n-1}. En la n-ésima ronda del juego, ocurren tres hechos en el siguiente orden:

(i) El conejo se mueve de forma invisible a un punto A_n tal que la distancia entre A_{n-1} y A_n es exactamente 1.

(ii) Un dispositivo de rastreo reporta un punto P_n al cazador. La única información segura que da el dispositivo al cazador es que la distancia entre P_n y A_n es menor o igual que 1.

(iii) El cazador se mueve de forma visible a un punto B_n tal que la distancia entre B_{n-1} y B_n es exactamente 1.

¿Es siempre posible que, cualquiera que sea la manera en que se mueva el conejo y cualesquiera que sean los puntos que reporte el dispositivo de rastreo, el cazador pueda escoger sus movimientos de modo que después de 10^9 rondas el cazador pueda garantizar que la distancia entre él mismo y el conejo sea menor o igual que 100?

Día 2 (19 de Julio, 2017)

Problema 4 (Por Charles Leytem, Luxemburgo)

Sean R y S puntos distintos sobre la circunferencia Ω tal que RS no es un diámetro de Ω. Sea ℓ la recta tangente a Ω en R. El punto T es tal que S es el punto medio del segmento RT. El punto J se elige en el menor arco RS de Ω de manera que Γ, la circunferencia circunscrita al triángulo JST, interseca a ℓ en dos puntos distintos. Sea A el punto común de Γ y ℓ más cercano a R. La recta AJ corta por segunda vez a Ω en K. Demostrar que la recta KT es tangente a Γ.

Problema 5 (Por Grigory Chelnokov, Russia)

Sea $N \geq 2$ un entero dado. Los $N(N+1)$ jugadores de un grupo de futbolistas, todos de distinta estatura, se colocan en fila. El técnico desea quitar $N(N-1)$ jugadores de esta fila, de modo que la fila resultante formada por los $2N$ jugadores restantes satisfaga las N condiciones siguientes:

(1) Que no quede nadie ubicado entre los dos jugadores más altos.

(2) Que no quede nadie ubicado entre el tercer jugador más alto y el cuarto jugador más alto.

...

(N) Que no quede nadie ubicado entre los dos jugadores de menor estatura.

Demostrar que esto es siempre posible.

Problema 6 (Por John Berman, Estados Unidos)

Un par ordenado (x, y) de enteros es un *punto primitivo* si el máximo común divisor de x e y es 1. Dado un conjunto finito S de puntos primitivos, demostrar que existe un entero positivo n y enteros a_0, $a_1, \ldots, a_n$, tal que para cada (x, y) de S, se cumple:

$$a_0 x^n + a_1 x^{n-1} y + a_2 x^{n-2} y^2 + \cdots + a_{n-1} x y^{n-1} + a_n y^n = 1.$$

IMO 2016

57° Olimpiada Internacional de Matemáticas

Hong Kong, Hong Kong | 06 – 16 de Julio, 2016.

Día 1 (11 de Julio, 2016)

Problema 1 (Por Art Waeterschoot, Bélgica)

El triángulo BCF es rectángulo en B. Sea A el punto de la recta CF tal que $FA = FB$ y F está entre A y C. Se elige el punto D de modo que $DA = DC$ y AC es bisectriz del ángulo $\angle DAB$. Asimismo, se elige el punto E de modo que $EA = ED$ y AD es bisectriz del ángulo $\angle EAC$. Así también, sea M el punto medio de CF y X un punto tal que $AMXE$ es un paralelogramo (con $AM \parallel EX$ y $AE \parallel MX$). Demostrar que las rectas BD, FX, y ME son concurrentes.

Problema 2 (Por Trevor Tao, Australia)

Hallar todos los enteros positivos n de modo que en cada casilla de un tablero de $n \times n$ se puede escribir una de las letras I, M y O tal que:

• en cada fila y en cada columna, un tercio de las casillas tiene I, un tercio tiene M y un tercio tiene O; y

• en cualquier línea diagonal compuesta por un número de casillas divisible por 3, exactamente un tercio de las casillas tienen I, un tercio tiene M y un tercio tiene O.

Nota: Las filas y las columnas del tablero de $n \times n$ se enumeran desde 1 hasta n, en su orden natural. Así, cada casilla corresponde a un par de enteros positivos (i, j) con $1 \le i, j \le n$. Para $n > 1$, el tablero tiene $4n - 2$ líneas diagonales de dos tipos. Una línea diagonal del primer tipo se compone de todas las casillas (i, j) para las que $i + j$ es una constante, mientras que una línea diagonal del segundo tipo se compone de todas las casillas (i, j) para las que $i - j$ es una constante.

Problema 3 (Por Aleksandr Gaifullin, Rusia)

Sea $P = A_1 A_2 \ldots A_k$ un polígono convexo en el plano. Los vértices A_1, A_2, ..., A_k tienen coordenadas enteras y se encuentran sobre una circunferencia. Sea S el área de P. Sea n un entero positivo impar tal que los cuadrados de las longitudes de los lados de P son todos números enteros divisibles por n. Demostrar que $2S$ es un entero divisible por n.

Dia 2 (12 de Julio, 2016)

Problema 4 (Por Gerhard Woeginger, Luxemburgo)

Un conjunto de números enteros positivos se llama *fragante* si contiene al menos dos elementos, y cada uno de sus elementos tiene algún factor primo en común con al menos uno de los elementos restantes. Sea $P(n) = n^2 + n + 1$. Determinar el menor número entero positivo b para el cual existe algún número entero no negativo a tal que el conjunto $\{P(a + 1), P(a + 2), \ldots, P(a + b)\}$ es fragante.

Problema 5 (Por Nazar Agakhanov y Ilya Bogdanov, Rusia)

En una pizarra se encuentra escrita la ecuación

$$(x - 1)(x - 2) \cdots (x - 2016) = (x - 1)(x - 2) \cdots (x - 2016)$$

que posee 2016 factores lineales en cada lado. Determinar el menor valor posible de k para el cual pueden borrarse exactamente k de estos 4032 factores lineales, de modo que al menos quede un factor en cada lado y la ecuación que resulte no tenga soluciones reales.

Problema 6 (Por Josef Tkadlec, República Checa)

Se tienen $n \geq 2$ segmentos en el plano tal que cada par de segmentos se intersecan en un punto, y no hay tres segmentos que tengan un punto en común. Jeff debe elegir uno de los extremos de cada segmento y colocar sobre él una rana mirando hacia el otro extremo. Luego silbará $n - 1$ veces. En cada silbido, cada rana saltará inmediatamente hacia adelante hasta el siguiente punto de intersección sobre su segmento. Las ranas nunca cambian las direcciones de sus saltos. Jeff quiere colocar las ranas de tal forma que nunca dos de ellas ocupen al mismo tiempo el mismo punto de intersección.

(a) Demostrar que si n es impar, Jeff siempre puede lograr su objetivo.

(b) Demostrar que si n es par, Jeff nunca logrará su objetivo.

IMO 2015

56° Olimpiada Internacional de Matemáticas

Chiang Mai, Tailandia | 04 – 16 de Julio, 2015.

Día 1 (10 de Julio, 2015)

Problema 1 (Por Merlijn Staps, Países Bajos)

Decimos que un conjunto finito de puntos S del plano es *equilibrado* si, para cada par de puntos diferentes A y B en S existe un punto C en S tal que $AC = BC$. Asimismo, decimos que S es *libre de centros* si, para cada terna de puntos distintos A, B y C en S no existe ningún punto P en S tal que $PA = PB = PC$.

(a) Demostrar que para todo $n \geq 3$ existe un conjunto de n puntos equilibrado.

(b) Determinar todos los enteros $n \geq 3$ para los que existe un conjunto de n puntos equilibrado y libre de centros.

Problema 2 (Por Dušan Djukić, Serbia)

Determinar todas las ternas (a, b, c) de enteros positivos tales que cada uno de los números $ab - c$, $bc - a$, $ca - b$ es una potencia de 2. *(Una potencia de 2 es entero de la forma 2^n, donde n es un entero no negativo)*

Problema 3 (Por Danylo Khilko y Mykhailo Plotnikov, Ucrania)

Sea ABC un triángulo acutángulo con $AB > AC$. Sea Γ su circunferencia circunscrita, H su ortocentro, y F el pie de la altura desde A. Sea M el punto medio del segmento BC. Sea Q el punto de Γ tal que $\angle HQA = 90°$ y sea K el punto de Γ tal que $\angle HKQ = 90°$. Supongamos que los puntos A, B, C, K y Q son todos distintos y están sobre Γ en este orden.

Demostrar que la circunferencia circunscrita al triángulo KQH es tangente a la circunferencia circunscrita al triángulo FKM.

Día 2 (11 de Julio, 2015)

Problema 4 (Por Silouanos Brazitikos y Evangelos Psychas, Grecia)

El triángulo ABC tiene la circunferencia circunscrita Ω con circuncentro O. Una circunferencia Γ de centro A corta al segmento BC en los puntos D y E tales que B, D, E y C son todos diferentes y están en la recta BC en ese orden. Sean F y G los puntos de intersección de Γ y Ω, tales que A, F, B, C y G están sobre Ω en este orden. Sea K el segundo punto de intersección de la circunferencia circunscrita al triángulo BDF y el segmento AB. Sea L el segundo punto de intersección de la circunferencia circunscrita al triángulo CGE y el segmento CA. Supongamos que las rectas FK y GL son distintas y se cortan en el punto X. Demostrar que X está sobre la recta AO.

Problema 5 (Por Dorlir Ahmeti, Albania)

Sea $\mathbb{R}$ el conjunto de los números reales. Determinar todas las funciones $f : \mathbb{R} \rightarrow \mathbb{R}$ que satisfacen la ecuación funcional

$$f(x + f(x + y)) + f(xy) = x + f(x + y) + yf(x)$$

para todos los números reales x, y.

Problema 6 (Por Ross Atkins e Ivan Guo, Australia)

La sucesión de enteros $a_1, a_2, \ldots$ satisface las siguientes condiciones:
(i) $1 \leq a_j \leq 2015$ para todo $j \geq 1$;
(ii) $k + a_k \neq \ell + a_\ell$ para todo $1 \leq k < \ell$.
Demostrar que existen dos enteros positivos b y N tales que

$$\left| \sum_{j=m+1}^{n} (a_j - b) \right| \leq 1007^2$$

para todos los enteros m y n que satisfacen $n > m \geq N$.

IMO 2014

55° Olimpiada Internacional de Matemáticas

Ciudad del Cabo, Sudáfrica | 03 – 13 de Julio, 2014.

Día 1 (08 de Julio, 2014)

Problema 1 (Por Gerhard Woeginger, Austria)

Sea $a_0 < a_1 < a_2 < \cdots$ una sucesión infinita de números enteros positivos. Demostrar que existe un único entero $n \geq 1$ tal que

$$a_n < \frac{a_0 + a_1 + \cdots + a_n}{n} \leq a_{n+1}. \quad (*)$$

Problema 2 (Por Tonči Kokan, Croacia)

Sea $n \geq 2$ un entero. Consideremos un tablero de tamaño $n \times n$ formado por n^2 cuadrados unitarios. Una configuración de n fichas en este tablero se dice que es *pacífica* si en cada fila y en cada columna hay exactamente una ficha. Hallar el mayor entero positivo k tal que, para cada configuración pacífica de n fichas, existe un cuadrado de tamaño $k \times k$ sin fichas en sus k^2 cuadrados unitarios.

Problema 3 (Por Ali Zamani, Irán)

En el cuadrilátero convexo $ABCD$, se tiene $\angle ABC = \angle CDA = 90°$. La perpendicular a BD desde A corta a BD en el punto H. Los puntos S y T están en los lados AB y AD, respectivamente, y son tales que H está dentro del triángulo SCT y

$$\angle CHS - \angle CSB = 90° \quad , \quad \angle THC - \angle DTC = 90°.$$

Demostrar que la recta BD es tangente a la circunferencia circunscrita del triángulo TSH.

Problema 4 (Por Giorgi Arabidze, Georgia)

Los puntos P y Q están en el lado BC del triángulo acutángulo ABC de modo que $\angle PAB = \angle BCA$ y $\angle CAQ = \angle ABC$. Los puntos M y N están en las rectas AP y AQ, respectivamente, de modo que P es el punto medio de AM, y Q es el punto medio de AN. Demostrar que las rectas BM y CN se cortan en la circunferencia circunscrita del triángulo ABC.

Problema 5 (Por Gerhard Woeginger, Luxemburgo)

Para cada entero positivo n, el Banco de Ciudad del Cabo produce monedas de valor $1/n$. Dada una colección finita de tales monedas (no necesariamente de distintos valores) cuyo valor total no supera $99 + 1/2$, demostrar que es posible separar esta colección en 100 o menos montones, de modo que el valor total de cada montón sea como máximo 1.

Problema 6 (Por Gerhard Woeginger, Austria)

Un conjunto de rectas en el plano está en *posición general* si no hay dos que sean paralelas ni tres que pasen por el mismo punto. Un conjunto de rectas en posición general separa el plano en regiones, algunas de las cuales tienen área finita; a estas las llamamos sus *regiones finitas*.

Demostrar que para cada n suficientemente grande, en cualquier conjunto de n rectas en posición general, es posible colorear de azul al menos $\sqrt{n}$ de ellas de tal manera que ninguna de sus regiones finitas tenga todos los lados de su frontera azules.

Nota: A las soluciones que reemplacen $\sqrt{n}$ por $c\sqrt{n}$ se les otorgarán puntos dependiendo del valor de c.

IMO 2013

54° Olimpiada Internacional de Matemáticas

Santa Marta, Colombia | 18 – 28 de Julio, 2013.

Día 1 (23 de Julio, 2013)

Problema 1 (Por CSP Japonés, Japón)

Probar que para cualquier par de enteros positivos k y n, existen k enteros positivos $m_1, m_2, \ldots, m_k$ (no necesariamente diferentes) tal que

$$1 + \frac{2^k - 1}{n} = \left(1 + \frac{1}{m_1}\right)\left(1 + \frac{1}{m_2}\right) \cdots \left(1 + \frac{1}{m_k}\right)$$

Problema 2 (Por Ivan Guo, Australia)

En una configuración de 4027 puntos del plano, donde 2013 son rojos y 2014 azules, en el cual que no hay tres de ellos que sean colineales, se llama *colombiana*. Después de trazarse algunas rectas, el plano queda dividido en varias regiones. Una colección de rectas es *buena* si para una configuración colombiana se cumple las siguientes condiciones:

• ninguna recta pasa por ningún punto de la configuración;
• ninguna región contiene puntos de ambos colores.

Hallar el mínimo valor de k tal que para cualquier configuración colombiana de 4027 puntos, exista una colección buena de k rectas.

Problema 3 (Por Alexander A. Polyansky, Rusia)

Asumamos que el excírculo del triángulo ABC opuesto al vértice A es tangente al lado BC en el punto A_1. Análogamente, se definen los puntos B_1 en CA y C_1 en AB, considerando los excírculos opuestos a

B y C respectivamente. Asimismo, asumamos que el circuncentro del triángulo $A_1B_1C_1$ se halla sobre la circunferencia que pasa por los vértices A, B y C. Probar que el triángulo ABC es rectángulo.

(El excírculo del triángulo ABC opuesto al vértice A es la circunferencia que es tangente al lado BC, y a la prolongación del lado AB más allá de B, y a la prolongación del lado AC más allá de C. Similarmente se definen los excírculos opuestos a los vértices B y C)

Día 2 (24 de Julio, 2013)

Problema 4 (Por W. Suksompong y P. Suteparuk, Tailandia)

Sea ABC un triángulo acutángulo con ortocentro H, y sea W un punto sobre el lado BC, situado entre B y C. Los puntos M y N son los pies de las alturas trazadas desde B y C, respectivamente. Se denota con ω_1 la circunferencia circunscrita al triángulo BWN, y con X al punto de ω_1 tal que WX es un diámetro de ω_1. En forma similar, se denota con ω_2 la circunferencia circunscrita al triángulo CWM, y con Y al punto de ω_2 tal que WY es un diámetro de ω_2. Probar que los puntos X, Y y H son colineales.

Problema 5 (Por Nikolai Nikolov, Bulgaria)

Let $\mathbb{Q}_{>0}$ be the set of rational numbers greater than zero. And let $f :$ $\mathbb{Q}_{>0} \to \mathbb{R}$ be a function Sea $\mathbb{Q}_{>0}$ el conjunto de los números racionales mayores a cero. Y sea $f : \mathbb{Q}_{>0} \to \mathbb{R}$ una función que satisface las siguientes condiciones:

(i) para todo $x, y \in \mathbb{Q}_{>0}$, $f(x)f(y) \geq f(xy)$;
(ii) para todo $x, y \in \mathbb{Q}_{>0}$, $f(x + y) \geq f(x) + f(y)$;
(iii) existe un número racional $a > 1$ tal que $f(a) = a$.
Probar que $f(x) = x$ para todo $x \in \mathbb{Q}_{>0}$.

Problema 6 (Por A. S. Golovanov y M. A. Ivanov, Rusia)

Sea n un número entero tal que $n \geq 3$. Consideremos ahora una circunferencia en donde se han marcado $n + 1$ puntos igualmente espaciados. Cada punto se etiqueta con uno de los números $0, 1, \ldots, n$ de manera que cada número sea usado exactamente una vez. Dos etiquetados se consideran el mismo, si uno de ellos se puede obtener del otro por una rotación de la circunferencia. Un etiquetado se llama *bonito*, si para cualesquiera cuatro etiquetas $a < b < c < d$ con $a + d = b + c$, la cuerda que une los puntos etiquetados con a y d no corta la cuerda que une los puntos etiquetados con b y c.

Sea M el número de etiquetados bonitos y N el número de pares ordenados (x, y) de enteros positivos tales que $x + y \leq n$ y $\mathrm{mcd}\,(x, y) = 1$. Probar que $M = N + 1$.

IMO 2012

53° Olimpiada Internacional de Matemáticas

Mar del Plata, Argentina | 04 – 16 de Julio, 2012.

Día 1 (10 de Julio, 2012)

Problema 1 (Por Evangelos Psychas, Grecia)

Dado un triángulo ABC, el punto J es el centro del excírculo opuesto al vértice A. Este excírculo es tangente al lado BC en M, y a las rectas AB y AC en K y L, respectivamente. Las rectas LM y BJ se cortan en F, y las rectas KM y CJ se cortan en G. Sea S el punto de intersección de las rectas AF y BC, y sea T el punto de intersección de las rectas AG y BC. Demostrar que M es el punto medio de ST.

(El excírculo de ABC opuesto al vértice A es la circunferencia que es tangente al lado BC, a la prolongación del lado AB más allá de B, y a la prolongación del lado AC más allá de C.)

Problema 2 (Por Angelo di Pasquale, Australia)

Sea $n \geq 3$ un entero, y sean $a_2, a_3, \ldots, a_n$ números reales positivos tales que $a_2 \cdot a_3 \cdot \ldots \cdot a_n = 1$. Probar que

$$(1 + a_2)^2(1 + a_3)^3 \cdots (1 + a_n)^n > n^n.$$

Problema 3 (Por David Arthur, Canadá)

El *juego de la adivinanza del mentiroso* es un juego que se realiza entre dos jugadores A y B. Las reglas del juego dependen de dos enteros positivos k y n que son conocidos por ambos jugadores. Al principio del juego, el jugador A elige enteros x y N con $1 \leq x \leq N$. El jugador A mantiene x en secreto, y le revela a B el valor real de N. A continuación, el jugador B trata de obtener información acerca de x

105

formulando preguntas a A de la siguiente manera: en cada pregunta, B especifica un conjunto arbitrario S de enteros positivos (que puede ser uno de los especificados en alguna pregunta anterior), y le pregunta a A si x pertenece a S. El jugador B puede hacer tantas preguntas de ese tipo como desee. Después de cada pregunta, el jugador A debe responderla inmediatamente con sí o no, pero puede mentir tantas veces como quiera. La única restricción es que entre cualesquiera $k+1$ respuestas consecutivas, al menos una debe ser cierta. Después de que B haya formulado tantas preguntas como haya deseado, debe especificar un conjunto X de a lo más n enteros positivos. Si x pertenece a X luego gana B; de lo contrario, pierde. Probar que:

(1) Si $n \geq 2^k$, entonces B puede garantizarse la victoria, y

(2) Para todo k suficientemente grande, existe un entero $n \geq 1.99^k$ tal que B no puede garantizarse la victoria.

Día 2 (11 de Julio, 2012)

Problema 4 (Por Liam Baker, Sudáfrica)
Hallar todas las funciones $f : \mathbb{Z} \to \mathbb{Z}$ que verifican la siguiente igualdad:

$$f(a)^2 + f(b)^2 + f(c)^2 = 2f(a)f(b) + 2f(b)f(c) + 2f(c)f(a),$$

para todos los enteros a, b, c donde $a + b + c = 0$. ($\mathbb{Z}$ denota el conjunto de los números enteros.)

Problema 5 (Por Josef Tkadlec, República Checa)
Sea ABC un triángulo tal que $\angle BCA = 90°$, y sea D el pie de la altura desde C. Sea X un punto interior del segmento CD. Sea K el punto en el segmento AX tal que $BK = BC$. En forma similar, sea L el punto en el segmento BX tal que $AL = AC$. Sea M el punto de intersección de AL y BK. Probar que $MK = ML$.

106

Problema 6 (Por Dušan Djukić, Serbia)

Hallar todos los enteros positivos n para los cuales existen enteros no negativos $a_1, a_2, \ldots, a_n$ tal que

$$\frac{1}{2^{a_1}} + \frac{1}{2^{a_2}} + \cdots + \frac{1}{2^{a_n}} = \frac{1}{3^{a_1}} + \frac{2}{3^{a_2}} + \cdots + \frac{n}{3^{a_n}} = 1.$$

IMO 2011

52° Olimpiada Internacional de Matemáticas

Amsterdam, Países Bajos | 12 – 24 de Julio, 2011.

Día 1 (18 de Julio, 2011)

Problema 1 (Por Fernando Campos, México)

Dado un conjunto $A = \{a_1, a_2, a_3, a_4\}$ de cuatro enteros positivos distintos se denota con S_A a la suma $a_1 + a_2 + a_3 + a_4$. Sea n_A el número de parejas (i, j) tal que $1 \leq i < j \leq 4$, de forma que $a_i + a_j$ divide a S_A. Hallar todos los conjuntos A de cuatro enteros positivos distintos de modo que se alcance el mayor valor posible de n_A.

Problema 2 (Por Geoff Smith, Reino Unido)

Sea S un conjunto finito de al menos dos puntos en el plano. Asimismo, en S no existen tres puntos colineales. Un *remolino* es un proceso que comienza con una recta ℓ que pasa por un único punto P de S. La recta ℓ se rota en el sentido de las manecillas del reloj con centro en P hasta que la recta encuentre por primera vez otro punto de S, al cual llamaremos Q. Con Q como nuevo centro, se sigue rotando la recta en el sentido de las manecillas del reloj hasta que la recta encuentre otro punto de S. Este proceso continúa indefinidamente. Demostrar que se puede elegir un punto P de S y una recta ℓ que pase por P tal que el remolino resultante use cada punto de S como centro de rotación un número infinito de veces.

Problema 3 (Por Igor Voronovich, Bielorrusia)

Sea $f: \mathbb{R} \longrightarrow \mathbb{R}$ una función definida en el conjunto de los números reales, la cual satisface la siguiente expresión

$$f(x + y) \le y\, f(x) + f(f(x))$$

para todo par de números reales x, y. Probar que $f(x) = 0$, $\forall\, x \le 0$.

Día 2 (19 de Julio, 2011)

Problema 4 (Por Morteza Saghafiyan, Irán)

Sea n un entero tal que $n > 0$. Se dispone de una balanza de dos platillos y de n pesas cuyos pesos son $2^0, 2^1, \ldots, 2^{n-1}$. Debemos colocar cada una de las n pesas en la balanza, una tras otra, de forma tal que el platillo derecho nunca sea más pesado que el platillo izquierdo. En cada paso, elegimos una de las pesas que no ha sido colocada en la balanza, y la colocamos ya sea en el platillo izquierdo o en el platillo derecho, hasta que todas las pesas hayan sido colocadas. Determinar el número de maneras en la que esto pueda realizarse.

Problema 5 (Por Seyedmahyar Sefidgaran, Irán)

Sea $f\colon \mathbb{Z} \longrightarrow \mathbb{Z}^+$, una función del conjunto de los enteros al conjunto de los enteros positivos. Asimismo, para cualesquiera dos enteros m y n, la diferencia $f(m) - f(n)$ es divisible por $f(m - n)$.

Probar que para todos los enteros m y n con $f(m) \le f(n)$, el número $f(n)$ es divisible por $f(m)$.

Problema 6 (Por CSP Japonés, Japón)

Sea ABC un triángulo acutángulo cuya circunferencia circunscrita es Γ. Sea ℓ una recta tangente a Γ, y sean ℓ_a, ℓ_b y ℓ_c las rectas las cuales se obtienen por reflexión de ℓ con respecto a las rectas BC, CA y AB, respectivamente. Demostrar que la circunferencia circunscrita al triángulo determinado por las rectas ℓ_a, ℓ_b y ℓ_c es tangente a la circunferencia Γ.

IMO 2010

51° Olimpiada Internacional de Matemáticas

Astaná, Kazajistán | 02 – 14 de Julio, 2010.

Día 1 (07 de Julio, 2010)

Problema 1 (Por Pierre Bornsztein, Francia)

Determine todas las funciones $f: \mathbb{R} \to \mathbb{R}$ tal que la igualdad

$$f(\lfloor x \rfloor y) = f(x) \lfloor f(y) \rfloor$$

se satisfaga para todos los números $x, y \in R$. ($\lfloor z \rfloor$ denota el mayor entero que es menor o igual a z.)

Problema 2 (Por Tai Wai Ming y Wang Chongli, Hong Kong)

Dado un triángulo ABC, sea I su incentro y Γ su circunferencia circunscrita. La recta AI interseca de nuevo a Γ en D. Sea E un punto en el arco $\overset{\frown}{BDC}$ y F un punto del lado BC tal que

$$\angle BAF = \angle CAE < \frac{1}{2}\angle BAC.$$

Asimismo, G es el punto medio del segmento IF. Demostrar que las rectas DG y EI se intersecan en Γ.

Problema 3 (Por Gabriel Carroll, Estados Unidos)

Sea $\mathbb{N}$ el conjunto de los enteros positivos. Determinar todas las funciones $g: \mathbb{N} \to \mathbb{N}$ tal que $(g(m) + n)(m + g(n))$ sea un cuadrado perfecto para todo $m, n \in \mathbb{N}$.

Problema 4 (Por Marcin Kuczma, Polonia)

Dado el triángulo ABC, sea Γ su circunferencia circunscrita y P un punto en su interior. Las rectas AP, BP y CP intersecan de nuevo a Γ en los puntos K, L y M, respectivamente. La recta tangente a Γ en C interseca a la recta AB en S. Si se cumple que $SC = SP$, probar que $MK = ML$.

Problema 5 (Por Hans Zantema, Países Bajos)

En cada una de las seis cajas $B_1, B_2, B_3, B_4, B_5, B_6$ hay inicialmente una sola moneda. Se permiten dos tipos de operaciones:

Tipo 1: Elegir una caja no vacía B_j, con $1 \leq j \leq 5$. Retirar una moneda de B_j y añadir dos monedas a B_{j+1}.

Tipo 2: Elegir una caja no vacía B_k, con $1 \leq k \leq 4$. Retirar una moneda de B_k e intercambiar los contenidos de las cajas (posiblemente vacías) B_{k+1} y B_{k+2}.

Determinar si existe una sucesión finita de estas operaciones que deje a las cajas B_1, B_2, B_3, B_4, B_5 vacías y a la caja B_6 conteniendo exactamente $2010^{2010^{2010}}$ monedas. (Notar que $a^{b^c} = a^{(b^c)}$.)

Problema 6 (Por Morteza Saghafiyan, Irán)

Sea $a_1, a_2, a_3, \ldots$ una sucesión de números reales positivos. Se tiene que para cierto entero positivo s,

$$a_n = max\{a_k + a_{n-k} \text{ tal que } 1 \leq k \leq n - 1\}$$

para todo $n > s$. Probar que existen enteros positivos ℓ y N, con $\ell \leq s$, tal que $a_n = a_\ell + a_{n-\ell}$ para todo $n \geq N$.